Advanced Water Injection for Low Permeability Reservoirs

Advanced Water Injection for Low Permeability Reservoirs

Editor

Sanjay Khanwelkar

scitus
academics

Advanced Water Injection for Low Permeability Reservoirs

Edited by **Sanjay Khanwelkar**

Printed in 2017

ISBN: 978-1-68117-330-6

Library of Congress Control Number: 2015939242

© 2016 by

SCITUS Academics LLC,
616, Corporate Way, Suite 2, 4766,
Valley Cottage, NY 10989

www.scitusacademics.com

Contents

Preface

Water Injection for Low Permeability Reservoirs provides operators with the proper workflow systems and engineering techniques for designing, planning and implementing water injection systems that will improve recovery factors. When used in low permeability or ultra-low permeability reservoirs, water injection is one of the most economical methods for ensuring maximum production rates. This book provides both theoretical analysis and practical cases for designing and evaluating water injection systems and understanding key production variables involved in making detailed predictions for oil and water producing rates, water injection rates, and recovery efficiency. This book clearly explains the characteristics of ultra-low permeability reservoirs and linear flow theories. These topics are then applied to design and implementation. Application cases of four oilfields are included to help develop concepts while illustrating the proper workflow for ensuring waterflooding performance analysis and optimization.

Editor

Modeling of Partially Hydrolyzed Polyacrylamide-Hexamine-Hydroquinone Gel System Used for Profile Modification Jobs in the Oil Field

Upendra Singh Yadav and Vikas Mahto

Department of Petroleum Engineering, Indian School of Mines, Dhanbad 826004, India

ABSTRACT

The cross-linked polymer gel systems are being used increasingly to redirect or modify reservoir fluid movement in the vicinity of injection wells for the purpose of permeability/profile modification job in the oil field due to their high temperature stability and capability to provide rigid gel having high mechanical strength. In this study, a partially hydrolyzed polyacrylamide-hexamine-hydroquinonegel is used for the development of polymer gel system. The experimental investigation demonstrates that the gelation time varies with polymer and crosslinker concentration and the temperature. The mathematical model is developed with the help of gelation kinetics of polymer gel and using Arrhenius equation, which relates the gelation time with polymer, crosslinker concentrations, and temperature. The developed model is solved with the help of multivariate regression method. It is observed in this study that the theoretical values of gelation time have good agreement with the experimental values.

INTRODUCTION

The reservoir heterogeneity, permeability variations, and presence of fractures and fracture network in the reservoir are the main hurdles of the water flooding operations used to enhance the oil recovery from matured oil fields. Normally during water flooding, water sweeps through more permeable sections of the reservoir leaving back oil in low permeability channels leading to low oil recovery and early water breakthrough. This excessive water production from the producers leads to rise in handling and disposal costs and reduces the economic life of the well. The polymer gels block or reduce the permeability in high permeability channels and divert the injected water through the low permeability sections, which were not flooded or swept earlier, leading to improvement in oil recovery. This technique is known as profile modification or permeability modification technique and the proper application of this technique is essential for the success of water flooding projects in the oil fields [1–3].

Basically two types of polymers have been used for profile modification jobs. These are polyacrylamides with different degrees of hydrolysis and polysaccharide such as xanthan biopolymer. These are cross linked with inorganic and organic crosslinkers to produce a three-dimensional polymer structure of the gel [4–7]. The inorganic gel system based on the crosslinking of the carboxylate groups on the partially hydrolyzed polyacrylamide chain (PHPA) with a trivalent cation like Cr (III). This crosslinking is believed to rely on coordination covalent bonding. It should be mentioned that Cr (III) based polymer gel system were reported to be stable at temperatures up to 300°F [8–10]. The organically crosslinked gels developed using phenol and formaldehyde crosslinkers are thermally stable under harsh environmental conditions. But, these crosslinkers are not environment friendly. To overcome the toxicity issues associated with formaldehyde and phenol, the formaldehyde can be replaced with less toxic derivatives like hexamethylenetetramine, glyoxal, paraformaldehyde, acetaldehyde, propionaldehyde, butyraldehyde and so forth and phenol can be replaced with hydroquinone, resorcinol, pyrogallol, phenyl salicylate and so forth [11–13]. Other system is based on PEI crosslinker and a copolymer of acrylamide and t-butyl acrylate (PA-t-BA). PA-t-BA is a relatively low molecular-weight polymer is expected to provide rigid ringing gels. Polyethyleneimine (PEI) cross-linking a copolymer of acrylamide and tert-butylacrylate (PAtBA) as water shutoff gels has been widely used in recent years [14].

The gelation mechanism of organic crosslinkers is done by covalent bonding, which is much more stable than the ionic bonds. Organic crosslinkers were introduced to obtained gels that can remain stable over a wide temperature range. For high temperature (> 90°C) reservoirs, acrylamide-based copolymers with organic crosslinking agents, such as phenol-formaldehyde and its derivatives, can be used to form a gel with thermal stability, and the gelation time is adjustable [15, 16].

The different methods are presented in the literature for the determination of the physical and chemical properties of polymer gel system such as bottle testing method, sealed tube method, dynamic shear method (rheometer), and static shear method (viscometer). Out of these methods, the bottle testing method is most suitable, faster, and inexpensive method to study the gelation kinetics [17].

Several models have been reported in the literature to relate the gelation time with temperature. Jordan found that the gelation time of a specified system decreases as the temperature is increased. Plots of the logarithm of gelation time versus the reciprocal of the absolute reaction temperature were found to be linear for the systems studied. This correlation showed that the system apparently followed the Hurd and Letteron model which assumed rate dependence on only one species. Hurd and Letteron studied the effect of temperature on the formation of silicic acid gels and found empirically that plot of the logarithm of the gelation time versus the reciprocal of the absolute temperature linear. They developed an equation, similar to the Arrhenius equation, which relates gelation time and the temperature and support the linear relationship [15]. But very few or rare models relate the gelation time with polymer and crosslinker concentration and temperature in case of polymer gel system used in the oil fields. For the successful design of the profile modification job the mathematical model equation consists of polymer and crosslinkers concentration as well as temperature is essential.

The present paper involves bulk gelation studies of the polymer gelant prepared from partially hydrolyzed polyacrylamide (PHPA) and hexamine (HMTA)/hydroquinone (HQ). The gelation time determined from the bulk gelation studies at different temperatures is useful because its knowledge at reservoir temperature is required to find out the depth up to which the gel can be placed in the petroleum reservoir rock. The gelation time and gel strength mainly depend on polymer and crosslinker concentrations and reservoir temperature were also reported in this paper. On the basis of the kinetics of polymer gel system, the mathematical model for the gelation behavior is also proposed which relates the gelation time with polymer-crosslinkers concentrations and temperature. The theoretical values of gelation time obtained from the proposed model matches with the experimental data.

GELATION MECHANISM

The hexamine on hydrolysis yields formaldehyde which then combines with hydroquinone and form 2,3,5,6-tetramethylol hydroquinone. Further, partially hydrolyzed polyacrylamide reacts with 2,3,5,6-tetramethylol hydroquinone and forms three dimensional networks of polymer gel, the different steps of which are as follows.

Step I: In the initial step, hexamine hydrolyses to yield formaldehyde

$$(CH_2)_6N_4 + H_2O \longrightarrow 6OH-CH_2-OH + 4NH_3$$
$$\underset{\text{HMTA} \quad + \text{ Water}}{} \qquad \underset{\text{Methane diol}}{} \tag{1}$$

$$OH-CH_2-OH \longrightarrow H-CH=O + H_2O$$
$$\underset{\text{Formaldehyde } + \text{ water}}{} \tag{2}$$

Step II. Formaldehyde produced in Step I then react with hydroquinone to form a condensed structure known as 2,3,5,6-tetramethylolhydroquinone (see Scheme 1).

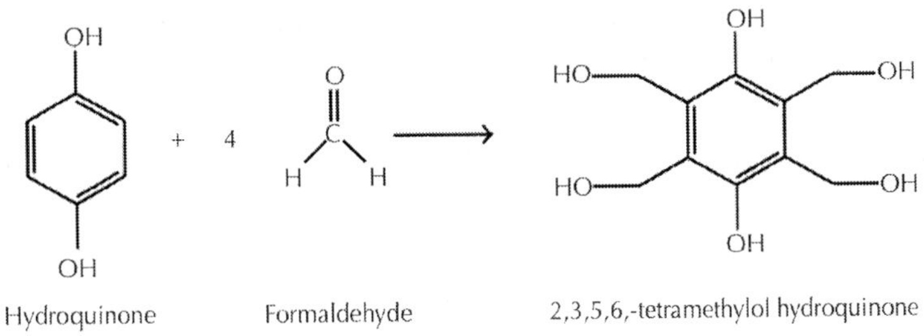
OH
OH
+ 4
O
C
H H
HO
OH
HO
OH
OH
OH
Hydroquinone
Formaldehyde
2,3,5,6,-tetramethylol hydroquinone

Scheme 1

Step III: The condensed molecule formed in Step II then reacts with PHPA to form the 3-dimensional gel structure which helps in profile modification jobs (see Scheme 2).

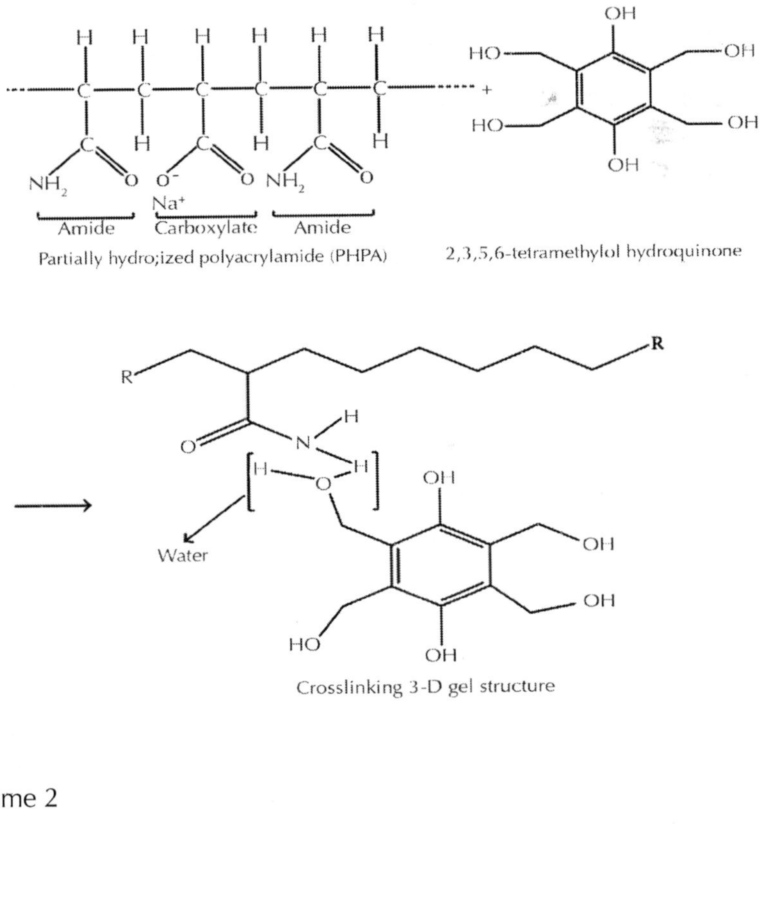
Amide
Carboxylate
Amide
Partially hydro;ized polyacrylamide (PHPA)
2,3,5,6-tetramethylol hydroquinone
Water
Crosslinking 3-D gel structure

Scheme 2

DEVELOPMENT OF MATHEMATICAL MODEL

The rate of gelation process for partially hydrolyzed polyacrylamide-hexamine-hydroquinone system may be present as follows.

$\alpha A + \beta B + \Upsilon C \longrightarrow 3D$ gel structure

$$r_A = -\frac{dC_A}{dt} = kC_A{}^a C_B{}^b C_C{}^c,$$

(3)

where k is the reaction rate constant, C_A, C_B, and C_c denotes the concentration of partially hexamine and hydroquinone and a, b, and c are the order with respect to A, B, and C, respectively. If the reactants are present in their stoichiometric ratios, they will remain at that ratio throughout the reaction [19]. Thus for reactants A, B, and C at any time,

$$\frac{C_B}{C_A} = \frac{\beta}{\alpha} \implies C_B = \frac{\beta}{\alpha}C_A, \qquad \frac{C_C}{C_A} = \frac{\gamma}{\alpha} \implies C_c = \frac{\gamma}{\alpha}C_A,$$

(4)

where α, β, and γ = reactant of species A, B, and C present in the reaction, respectively

$$r_A = -\frac{dC_A}{dt} = kC_A{}^a \cdot \left(\frac{\beta}{\alpha}C_A\right)^b \cdot \left(\frac{\gamma}{\alpha}C_A\right)^c$$

$$= k\left(\frac{\beta}{\alpha}\right)^b \left(\frac{\gamma}{\alpha}\right)^c \cdot C_A{}^{a+b+c}.$$

(5)

The above equation can be also written as

$$r_A = -\frac{dC_A}{dt} = k'C_A{}^n.$$

(6)

Taking integration both side w.r.t. time, we have

$$t = \frac{1}{(n-1)k'}\left\{\frac{1}{C_A{}^{n-1}} - \frac{1}{C_{Ao}{}^{n-1}}\right\}$$

(7)

Rearranging the above equation and putting the value of k' and n,

$$t = \frac{1}{k\{(a+b+c)-1\}} \left\{ \frac{C_A}{C_A{}^a \cdot C_A{}^b (\beta/\alpha)^b \cdot C_A{}^c (\gamma/\alpha)^c} \right.$$

$$\left. - \frac{C_{Ao}}{C_{Ao}{}^a \cdot C_{Ao}{}^b (\beta/\alpha)^b \cdot C_{Ao}{}^c (\gamma/\alpha)^c} \right\} \quad (8)$$

Rearranging the above equation, we get

$$t = \frac{1}{k\{(a+b+c)-1\}} \left\{ \frac{C_A}{C_A{}^a C_B{}^b C_C{}^c} - \frac{C_{Ao}}{C_{Ao}{}^a C_{Bo}{}^b C_{Co}{}^c} \right\}, \quad (9)$$

where C_{Ao}, C_{Bo}, and C_{Co} = initial concentrations of polymer, hexamine & hydroquinone.

By generalizing the above equation, it can be written as

$$t \propto \frac{1}{C_A{}^a C_B{}^b C_C{}^c}. \quad (10)$$

The effect of temperature (T in degree kelvin) can also be incorporated in (8) and Arrhenius model [19] can be utilized to connect k with T which is as follows:

$$k = A \exp\left(\frac{-E_a}{RT}\right), \quad (11)$$

where E_a is an apparent activation energy and R is the gas constant. Therefore, the combination of (10) and (11), then we get

$$t \propto \frac{1}{C_A{}^a C_B{}^b C_C{}^c} \times \exp\left(\frac{E_a}{RT}\right). \quad (12)$$

The above equation can be written as follows, k″ is the proportionality constant:

$$t = k'' \frac{1}{C_A{}^a C_B{}^b C_C{}^c} \times \exp\left(\frac{E_a}{RT}\right). \quad (13)$$

Taking natural logarithm and introducing the coefficients a, b, c then we get,

$$\ln t = \ln \left\{ k'' \frac{1}{C_A{}^a C_B{}^b C_C{}^c} \times \exp\left(\frac{E_a}{RT}\right) \right\},$$
(14)

$$\ln t = a_0 + a_1 \ln C_A + a_2 \ln C_B + a_3 \ln C_C + \frac{a_4}{T}.$$
(15)

Taking antilog both sides and assume t then we get

$$t = C_A{}^{a_1} \times C_B{}^{a_2} \times C_C{}^{a_3} \times \exp\left(a_0 + \frac{a_4}{T}\right).$$
(16)

Equation (16) is the general equation for the gelation behavior of partially hydrolyzed polyacrylamide-hexamine-hydroquinone gel system. The value of constant depends upon the polymer and crosslinkers composition and temperature.

SOLUTION PROCEDURE OF THE DEVELOPED MODEL

The mathematical model for gelation time consists of four variable parameters (C_A, C_B, C_C, and T) and five constants (a_0, a_1, a_2, a_3, and a_4). For the determination of these constants, five simultaneous equations are generated by multiplying the variables in both side of the model equation and utilizing the experimental data, the flow diagram for solution of proposed model is shown in Figure 1. Further, these equations are solved by multivariate regression method and values of the constants a_0, a_1, a_2, a_3, and a_4 are determined. After putting these values in the model equation, the gelation time are calculated by varying the different parameters (C_A, C_B, C_C, and T) finally, these calculated values are compared with the actual values obtained from the laboratory work.

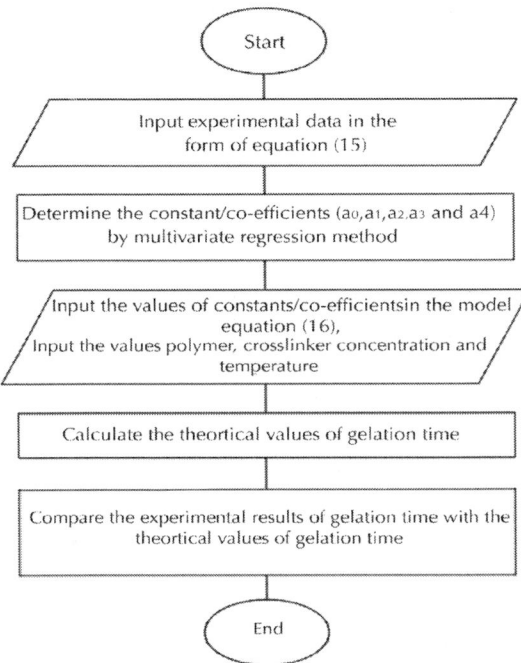

Figure 1: Flow chart for the solution of developed model.

EXPERIMENTAL WORK

Material Used

The materials used for this work are partially hydrolyzed polyacrylamide, hexamine, hydroquinone, sodium chloride, hydrochloric acid, and sodium hydroxide. Partially hydrolyzed polyacrylamide is procured from Oil and Natural Gas Corporation Limited, Mumbai, India. Hexamine is purchased from Otto Kemi Mumbai, India and hydroquinone is procured from Ranbaxy Fine Chemicals Ltd., New Delhi, India. Hydrochloric acid is purchased from Central Drug house (P) Ltd. New Delhi, India. Sodium chloride is purchased from Nice Chemical Pvt. Ltd. Cochin, India and sodium hydroxide is purchased from S. D. Fine-Chem Ltd. Mumbai, India.

Experimental Procedure

Initially stock solution of partially hydrolyzed polyacrylamide was prepared in brine solution and kept for aging at ambient temperature for 24 hrs and further fresh solution of hexamine and hydroquinone were also prepared in brine. The appropriate solution of partially hydrolyzed polyacrylamide, hexamine, hydroquinone, and brine were thoroughly mixed at room temperature by magnetic stirrer. The pH of the gelant solution was measured by Century CP-901 digital pH meter and the pH of the gelant solution was adjusted by using 1 N sodium hydroxide and 1 N hydrochloric acid. Finally, the gelant solution was transferred into small glass bottle and kept at the desire temperature in the hot air oven (temperature ranges from 80°C to 120°C). Due to the increased temperature and the crosslinking reaction, gel formation takes place. The quality of gel was visually inspected at regular intervals and gelation time was noted. Here, the time for the formation of stiff/rigid gel was only considered.

RESULTS AND DISCUSSIONS

The gelation behavior of the polymer gel system largely depends on the polymer-crosslinker concentrations and the temperature. The effects of these parameters on gelation time are described below.

Effect of Temperature on Gelation Time

Different gelling solutions were prepared with different concentrations of polymer and crosslinking agent at pH 7.5 and kept for gelation at different temperature. The reaction rate between amide group and methylol group was accelerated by increasing temperature and the gelation time decreased. Increasing the gelation temperature results in a decrease in gelation time since at higher temperature gels are formed in lesser time due to rapid crosslinking as is depicted in Figures 2 and 3 and Tables 1, 2, 3, 4, and 5. A possible explanation for rapid crosslinking is either due to an increase in molecular mobility or formation of new cross-linking sites as a result of gelation reaction. It is a known fact that degree of hydrolysis of the polymer increases at elevated temperatures which in turn increases the number of cross-linking sites that is found increasing the reaction rate and decreasing the gelation time.

Table 1: Experimental values of gelation time at different polymer concentration, temperature and pH 7.5 (Constant crosslinker concentration HMTA/HQ 0.5/0.4 wt%)

| Sl. No. | Polymer PHPA (wt%) | Crosslinkers | | Experimental values of gelation time (hrs.) | Theoretical values of gelation time (hrs.) | Temperature (C) |
		HMTA (wt%)	HQ (wt%)			
1	0.8	0.5	0.4	280	328.8475	80
2	0.9	0.5	0.4	200	178.3855	
3	1.0	0.5	0.4	120	103.3145	
4	1.1	0.5	0.4	61	62.9199	
5	0.8	0.5	0.4	174.3	163.9435	90
6	0.9	0.5	0.4	82.3	88.9322	
7	1.0	0.5	0.4	49	51.4565	
8	1.1	0.5	0.4	29.3	31.3681	
9	0.8	0.5	0.4	87	84.8392	100
10	0.9	0.5	0.4	41	46.0215	
11	1.0	0.5	0.4	31	26.6282	
12	1.1	0.5	0.4	15.3	16.2326	
13	0.8	0.5	0.4	45.3	45.4394	110
14	0.9	0.5	0.4	23	24.6489	
15	1.0	0.5	0.4	13	14.2619	
16	1.1	0.5	0.4	7	8.6941	
17	0.8	0.5	0.4	28	25.1225	120
18	0.9	0.5	0.4	14	13.6278	
19	1.0	0.5	0.4	8	6.8917	
20	1.1	0.5	0.4	5	4.80681	

Table 2: Experimental values of gelation time at different polymer concentration, temperature and pH 7.5 (Constant crosslinker concentration HMTA/HQ 0.5/0.3 wt%)

| Sl. No. | Polymer | Crosslinkers | | Experimental values of gelation time (hrs.) | Theoretical values of gelation time (hrs.) | Temperature (°C) |
	PHPA (wt%)	HMTA (wt%)	HQ (wt%)			
1	0.8	0.5	0.3	312	376.7173	80
2	0.9	0.5	0.3	230	204.3528	
3	1.0	0.5	0.3	134	118.2393	
4	1.1	0.5	0.3	68	72.0791	
5	0.8	0.5	0.3	207	187.8085	90
6	0.9	0.5	0.3	91	101.8779	
7	1.0	0.5	0.3	55.3	58.9469	
8	1.1	0.5	0.3	33.3	35.9343	
9	0.8	0.5	0.3	101	97.1891	100
10	0.9	0.5	0.3	55.3	52.7208	
11	1.0	0.5	0.3	34	30.5045	
12	1.1	0.5	0.3	19	18.5956	
13	0.8	0.5	0.3	50	52.0539	110
14	0.9	0.5	0.3	25.3	28.237	
15	1.0	0.5	0.3	14.3	16.338	
16	1.1	0.5	0.3	8	9.9597	
17	0.8	0.5	0.3	32	28.7795	120
18	0.9	0.5	0.3	14.3	15.6116	
19	1.0	0.5	0.3	8.3	9.0329	
20	1.1	0.5	0.3	6.3	5.5065	

Table 3: Experimental values of gelation time at different polymer concentration, temperature and pH 7.5 (Constant crosslinker concentration HMTA/HQ 0.4/0.4 wt %)

Sl. No.	Polymer	Crosslinkers		Experimental values of gelation time (hrs.)	Theoretical values of gelation time (hrs.)	Temperature (°C)
	PHPA (wt%)	**HMTA (wt%)**	**HQ (wt%)**			
1	0.8	0.4	0.4	331	411.5734	80
2	0.9	0.4	0.4	243	223.2606	
3	1.0	0.4	0.4	140	129.1795	
4	1.1	0.4	0.4	77	78.7483	
5	0.8	0.4	0.4	226	205.1856	90
6	0.9	0.4	0.4	105	111.3043	
7	1.0	0.4	0.4	71	64.4011	
8	1.1	0.4	0.4	37.3	39.2591	
9	0.8	0.4	0.4	115	106.1816	100
10	0.9	0.4	0.4	61	57.5989	
11	1.0	0.4	0.4	37	33.3269	
12	1.1	0.4	0.4	23	20.3162	
13	0.8	0.4	0.4	56	56.8703	110
14	0.9	0.4	0.4	30	30.8496	
15	1.0	0.4	0.4	17	17.8497	
16	1.1	0.4	0.4	10	10.8812	
17	0.8	0.4	0.4	36	31.4424	120
18	0.9	0.4	0.4	17	17.0561	
19	1.0	0.4	0.4	9	9.8687	
20	1.1	0.4	0.4	7	6.016	

Table 4: Experimental values of gelation time at different polymer concentration, temperature and pH 7.5 (Constant crosslinker concentration HMTA/HQ 0.4/0.3 wt %)

Sl. No.	Polymer	Crosslinkers		Experimental values of gelation time (hrs.)	Theoretical values of gelation tim (hrs.)	Temperature (°C)
	PHPA (wt %)	HMTA (wt %)	HQ (wt %)			
1	0.8	0.4	0.3	370	471.4855	80
2	0.9	0.4	0.3	286	255.7604	
3	1.0	0.4	0.3	168	147.9839	
4	1.1	0.4	0.3	84	90.2116	
5	0.8	0.4	0.3	266	235.0542	90
6	0.9	0.4	0.3	135.3	127.5067	
7	1.0	0.4	0.3	78	73.7758	
8	1.1	0.4	0.3	42.3	44.974	
9	0.8	0.4	0.3	133	121.6383	100
10	0.9	0.4	0.3	71	65.9835	
11	1.0	0.4	0.3	42	38.1783	
12	1.1	0.4	0.3	27	23.2736	
13	0.8	0.4	0.3	60	65.1488	110
14	0.9	0.4	0.3	35	35.3404	
15	1.0	0.4	0.3	20.3	20.4481	
16	1.1	0.4	0.3	12	12.4652	
17	0.8	0.4	0.3	39	36.0194	120
18	0.9	0.4	0.3	19	19.539	
19	1.0	0.4	0.3	10	11.3053	
20	1.1	0.4	0.3	8	6.8917	

Table 5: Experimental values of gelation time at different polymer concentration, temperature and pH 7.5 (Constant crosslinker concentration HMTA/HQ 0.3/0.3 wt %)

S l. No.	Polymer	Crosslinkers		Experimental values of gelation time (hrs.)	Theoretical values of gelation time (hrs.)	Temperature (°C)
	PHPA (wt%)	HMTA (wt%)	H Q (wt%)			
1	0.8	0.3	0.3	451	629.6609	80
2	0.9	0.3	0.3	362	341.5637	
3	1.0	0.3	0.3	216	197.63	
4	1.1	0.3	0.3	111	120.476	
5	0.8	0.3	0.3	311	313.9109	90
6	0.9	0.3	0.3	158	170.283	
7	1.0	0.3	0.3	102	98.5264	
8	1.1	0.3	0.3	55.3	60.0621	
9	0.8	0.3	0.3	170	162.4459	100
10	0.9	0.3	0.3	82	88.1198	
11	1.0	0.3	0.3	58	50.9864	
12	1.1	0.3	0.3	30	31.0815	
13	0.8	0.3	0.3	86	87.0051	110
14	0.9	0.3	0.3	66	57.1604	
15	1.0	0.3	0.3	24.3	27.308	
16	1.1	0.3	0.3	15	16.6471	
17	0.8	0.3	0.3	45	48.1034	120
18	0.9	0.3	0.3	23	26.094	
19	1.0	0.3	0.3	14	15.098	
20	1.1	0.3	0.3	10	9.2038	

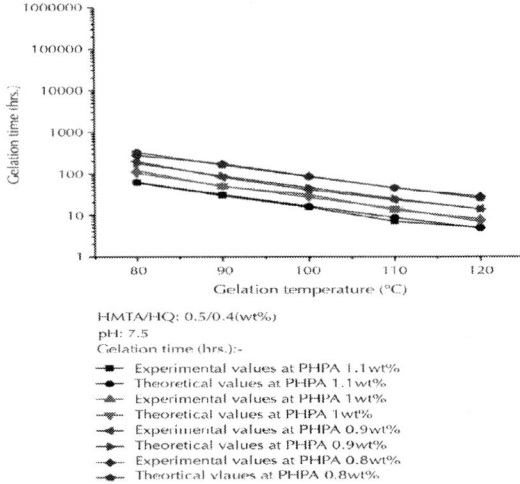

HMTA/HQ: 0.5/0.4(wt%)
pH: 7.5
Gelation time (hrs.):-

- ■ Experimental values at PHPA 1.1wt%
- ● Theoretical values at PHPA 1.1wt%
- ▲ Experimental values at PHPA 1wt%
- ▼ Theoretical values at PHPA 1wt%
- ◆ Experimental values at PHPA 0.9wt%
- ► Theoretical values at PHPA 0.9wt%
- ◆ Experimental values at PHPA 0.8wt%
- ● Theortical vlaues at PHPA 0.8wt%

Figure 2: Plot of gelation temperature versus experimental and theoretical values of gelation time at different polymer concentration, constant cross-linker concentration (HMTA/HQ 0.5/0.4 wt%), and pH 7.5.

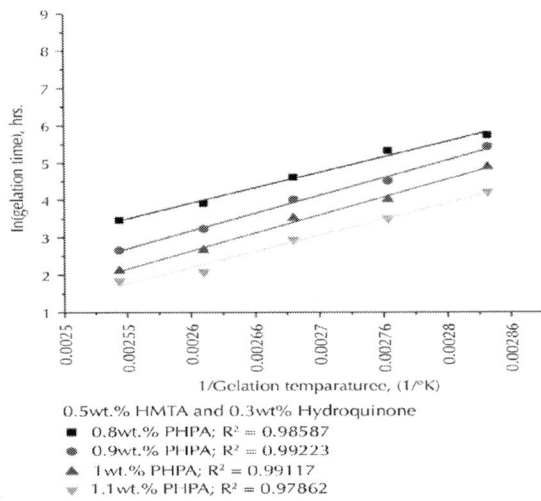

0.5wt.% HMTA and 0.3wt% Hydroquinone
- ■ 0.8wt.% PHPA; $R^2 = 0.98587$
- ● 0.9wt.% PHPA; $R^2 = 0.99223$
- ▲ 1wt.% PHPA; $R^2 = 0.99117$
- ▼ 1.1wt.% PHPA; $R^2 = 0.97862$

Figure 3: Plot for the temperature dependence of gelation time according to Arrhenius-type equation.

Effect of Polymer Concentration on Gelation Time

The polymer concentration has a significant effect on the physical properties of gel. The polymer concentration increases and gelation time decreases were shown in Figure 4. The polymer concentration ranges from 0.8 to 1.1 wt% and constant concentration of crosslinker (0.5 to 0.3 wt% HMTA and 0.4 to 0.3 wt% HQ) at pH 7.5 are depicted in Tables 1 to 5. As the polymer concentration increases, it means more crosslinking sites are available for the fast crosslinking reaction takes place. Thus, the gel formation reaction increases, which leads to the decreases of gelation time. This trend is expected to be the same at all gelation temperatures under study. Thus, the required time to obtain a non-flowing polymer gel with a tolerable strength decreased when the polymer concentration was increased.

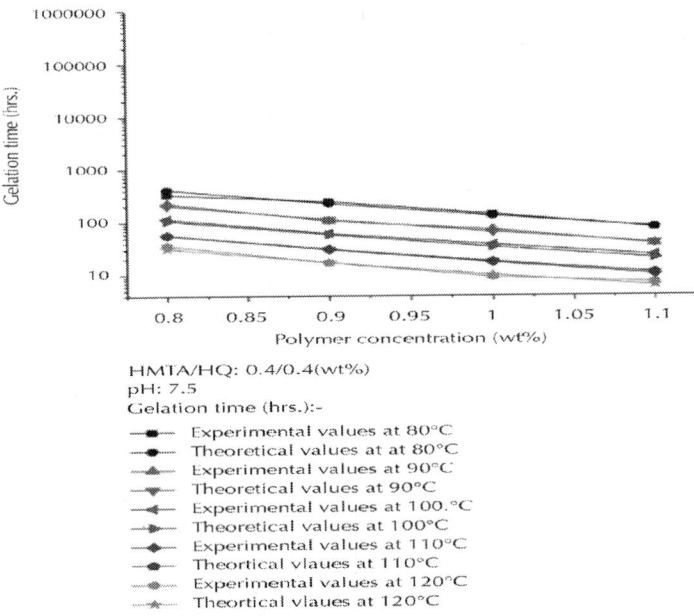

Figure 4: Plot of polymer concentration versus experimental and theoretical values of gelation time at different temperature, constant crosslinker concentration (HMTA/HQ 0.4/0.4 wt %) and pH 7.5.

Effect of Cross-Linker Concentration on Gelation Time

Hexamine and hydroquinone crosslinkers are a multifunctional group which can build a complex network with amide groups of partially hydrolyzed polyacrylamide and form a 3-dimentional gel network structure. Crosslinker concentration has a significant effect on the gel strength. The crosslinker concentration increases and gelation time decreases were shown in Figure 5. The several samples were prepared to investigate the effect of crosslinker concentration on the network strength. Bottle testing results shown in Tables 1 to 5 indicate that when the concentration of both crosslinker was decreased the gelation rate and gel quality is also decreased. In other words, when crosslinking agent concentration was increased, the stage of polymer gel changed from a state of flowing gel to one of deformable non-flowing gel, because crosslinking sites are increased for the formation of gel in lesser time intervals.

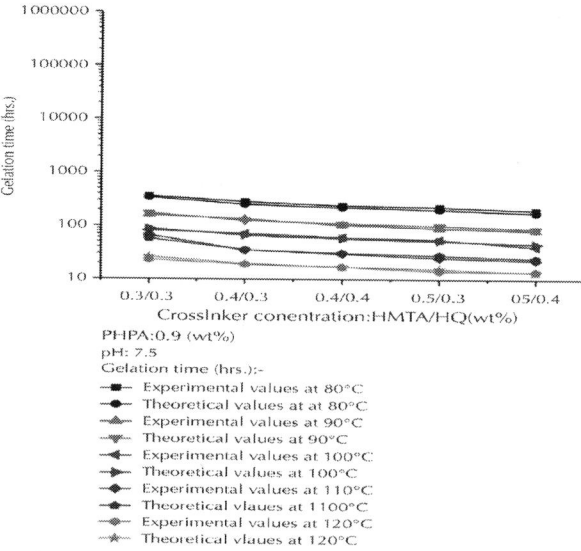

Figure 5: Plot of crosslinker concentration versus experimental and theoretical values of gelation time at different temperature, constant polymer concentration (PHPA 0.9 wt %), and pH 7.5.

MODEL DISCUSSION

Equation (12) shows the relation of gelation time with polymer concentration, cross linker concentrations and temperature. Here C_A, C_B, and C_C are polymers, HMTA and hydroquinone concentrations, respectively, in gm/litre and temperature, are taken in degree kelvin. This equation indicates that the gelation time is inversely proportional to the concentrations of polymer and crosslinking agent, which reflects that the increasing the concentration of polymer and crosslinking agent decreases the gelation time. The proposed model for the study of the gelation behavior of partially hydrolyzed polyacrylamide-hexamine-hydroquinone polymer gel system is shown in (15)-(16). The values of the constants/coefficients a_0, a_1, a_2, a_3, and a_4 are determined by multivariate regression method.

Equation (15) can be rewritten as follows:

$$y = a_0 + a_1 \cdot x_1 + a_2 \cdot x_2 + a_3 \cdot x_3 + a_4 \cdot x_4, \quad (17)$$

Where

$$y = \ln t; \qquad x_1 = \ln C_A; \qquad x_2 = \ln C_B,$$

$$x_3 = \ln C_C, \qquad x_4 = \frac{1}{T}. \quad (18)$$

The above equation solved by multivariate regression method and the values of constants a_0, a_1, a_2, a_3, and a_4 are calculated. The values of the constants a_0, a_1, a_2, a_3, and a_4 are found to be 1.77831, −5.19301, −1.00564, −0.47246 and 8926.87612, respectively and finally the gelation time is calculated from the following equation:

$$t = C_A^{-5.19301} \times C_B^{-1.00564} \times C_C^{-0.47246}$$

$$\times e^{(1.77831+(8926.87612/T))}. \quad (19)$$

The value of R^2 is 0.99122, which shows the good agreement between experimental and theoretical values of gelation time. This study reveals that the developed model may be used for the study of gelation behavior of partially hydrolyzed polyacrylamide-hexamine-hydroquinone gel system for its application in profile modification jobs. The theoretical values calculated from above equation are shown in Tables 1–5 at different polymer-crosslinker concentration and temperature.

Comparison between Developed Model and Civan et al. [18]

The developed model in the present study relates stoichiometric relation between reactants involved in the gelation reaction as well as Arrhenius equation. In this case, gelation time is dependent on polymer and crosslinker concentration as well as temperature. However, model proposed by Civan et al. is simply Arrhenius type equation which is dependent on temperature only. Figure 3 shows the theoretical values of gelation time for different concentration of polymers. It was found that there are several variations in the values of slopes and intercepts as well as R^2 and may have different activation energy for different concentrations of polymers. The comparison of theoretical gelation time calculated using Civan et al. model and our developed model at the concentration (0.8% polymer) are approximately the same which is shown in Table 6 and have good agreement with that particular experimental data. But, the developed model presents the same values of coefficients and activation energy for whole experimental study and all data are fitted in that mathematical equation. Hence, it may be more realistic in nature.

Table 6: Comparison of gelation time of developed model and Arrhenius type equation

Temperature (°C)	Experimental gelation time (hrs.)	Theoretical gelation time (hrs.) by Arrhenius type equation {Civan et al. [18]} $\ln(\square)=8288.47281/T(°K) -17.63044$ (0.8 wt% PHPA, 0.5 wt% HMTA and 0.3 wt% hydroquinone)	Theoretical gelation time (hrs.) by Developed Model (0.8 wt% PHPA, 0.5 wt% HMTA and 0.3 wt% hydroquinone)
80	312	343.66928	376.66426
90	207	180.07746	187.78207
100	101	97.68365	97.17542
110	50	54.70794	52.04664
120	32	31.55635	28.77554

CONCLUSIONS

The following conclusions are drawn from the present study.

- The gelation time of partially hydrolyzed polyacrylamide-hexamine-hydroquinone gel system decreases with increasing the cross linker concentration and temperature.

- On the basis of kinetics of gelation behavior the mathematical model for the partially hydrolyzed polyacrylamide-hexamine-hydroquinone gel system is the following:

$$t = C_A{}^{a_1} \times C_B{}^{a_2} \times C_C{}^{a_3} \times e^{(a_0 + (a_4/T))} \qquad (20)$$

- The values of constants a_0, a_1, a_2, a_3, and a_4 for this study are found to be 1.77831, −5.19301, −1.00564, −0.47246, and 8926.87612, respectively.

- The final mathematical model for the present study is as follows:

$$t = C_A{}^{-5.19301} \times C_B{}^{-1.00564} \times C_C{}^{-0.47246}$$
$$\times e^{(1.77831 + (8926.87612/T))} . \qquad (21)$$

- A values show good agreement between experimental and theoretical values of gelation time. This study reveals that the developed model may be used for the study of gelation behavior of partially hydrolyzed polyacrylamide-hexamine-hydroquinone gel system for its application in profile modification jobs.

ACKNOWLEDGMENTS

The authors are very thankful to Council of Scientific and Industrial Research (CSIR), New Delhi, India for financial assistance to carry out the research work.

REFERENCES

1. Y. Liu, B. Bai, and Y. Wang, "Applied technologies and prospects of conformance control treatments in China," Oil and Gas Science and Technology, vol. 65, no. 6, pp. 859–878, 2010. · ·

2. R. Jain, C. S. McCool, D. W. Green, G. P. Willhite, and M. J. Michnick, "Reaction kinetics of the uptake of chromium (III) acetate by polyacrylamide," Society of Petroleum Engineers Journal, vol. 10, no. 3, pp. 247–255, 2004. ·

3. C. A. Grattoni, H. H. Al-Sharji, C. Yang, A. H. Muggeridge, and R. W. Zimmerman, "Rheology and permeability of crosslinked polyacrylamide gel," Journal of Colloid and Interface Science, vol. 240, no. 2, pp. 601–607, 2001. · ·

4. A. Moradi-Araghi, "A review of thermally stable gels for fluid diversion in petroleum production,"Journal of Petroleum Science and Engineering, vol. 26, no. 1–4, pp. 1–10, 2000. · ·

5. A. Stavland and H. C. Jonsbraten, "New insight into aluminum citrate/polyacrylamide gels for fluid control," in Proceedings of the SPE/DOE 10th Symposium on Improved Oil Recovery, Paper SPE/DOE 35381, pp. 347–356, Tulsa, Okla, USA, April 1996. · ·

6. S. L. Bryant, G. P. Borghi, M. Bartosek, and T. P. Lockhart, "Experimental investigation on the injectivity of phenol-formaldehyde/polymer gelants," in Proceedings of the SPE International Symposium on Oilfield Chemistry, Paper SPE 37244, pp. 335–343, Houston, Tex, USA, February 1997. · ·

7. H. Jia, W. F. Pu, J. Z. Zhao, and R. Liao, "Experimental investigation of the novel phenol-formaldehyde cross-linking HPAM gel system: based on the secondary cross-linking method of organic cross-linkers and its gelation performance study after flowing through porous media," Energy and Fuels, vol. 25, no. 2, pp. 727–736, 2011.

8. H. Jia, W. F. Pu, J. Z. Zhao, and F. Y. Jin, "Research on the gelation performance of low toxic PEI cross-linking PHPAM gel systems as water shutoff agents in low temperature reservoirs," Industrial and Engineering Chemistry Research, vol. 49, no. 20, pp. 9618–9624, 2010. · ·

9. H. T. Dovan, R. D. Hutchins, and B. B. Sandiford, "Delaying gelation of aqueous polymers at elevated temperatures using novel organic crosslinkers," in Proceedings of the SPE International Symposium on Oilfield Chemistry, Paper SPE 37246, pp. 361–371, Houston, Tex, USA, February 1997.

10. R. D. Hutchins, H. T. Dovan, and B. B. Sandiford, "Field applications of high temperature organic gels for water control," in Proceedings of the 10th Symposium of Improved Oil Recovery, Paper SPE/DOE 35444, Tulsa, Okla, USA, April 1996.

11. L. Eoff, D. Dalrymple, and D. Everett, "Global field results of a polymeric gel system in conformance applications," in Proceedings of SPE Russian Oil and Gas Technical Conference and Exhibition, Paper SPE 101822, pp. 280–285, Moscow, Russia, October 2006. · ·

12. J. Vasquez, E. D. Dalrymple, L. Eoff, B. R. Reddy, and F. Civan, "Development and evaluation of high-temperature conformance polymer systems," in Proceedings of the SPE International Symposium on Oilfield Chemistry, Paper SPE 93156, Houston, Tex, USA, February 2005. · ·

13. G. A. Al-Muntasheri, H. A. Nasr-El-Din, and I. A. Hussein, "A rheological investigation of a high temperature organic gel used for water shut-off treatments," Journal of Petroleum Science and Engineering, vol. 59, no. 1-2, pp. 73–83, 2007.

14. G. A. Al-Muntasheri, "A study of polyacrylamide-based gels crosslinked with polyethyleneimine," inProceedings of the SPE International Symposium on Oilfield Chemistry, Paper SPE 105925, Houston, Tex, USA, February 2007. ·

15. S. D. Jordan, D. W. Green, E. R. Terry, and G. P. Willhite, "The effect of temperature on gelation time for polyacrylamide/ chromium (III) systems," Society of Petroleum Engineers Journal, vol. 22, no. 4, pp. 463–471, 1982. ·

16. G. A. Al-Muntasheri, H. A. Nasr-El-Din, and P. L. J. Zitha, "Gelation kinetic and performance evaluation of an organically crosslinked gel at high temperature and pressure," in Proceedings of the 1st International Oil Conference and Exhibition, Paper SPE 104071, September 2006, Cancun, Mexico. ·

17. M. Simjoo, M. Vafaie Sefti, A. Dadvand Koohi, R. Hasheminasab, and V. Sajadian, "Polyacrylamide gel polymer as water shut-off system: preparation and investigation of physical and chemical properties in one of the iranian oil reservoirs conditions," Iranian Journal of Chemistry and Chemical Engineering, vol. 26, no. 4, pp. 99–108, 2007.

18. F. Civan, J. Vasquez, D. Dalrymple, L. Eoff, and B. R. Reddy, "Laboratory and theoretical evaluation of gelation time data for water-based polymer systems for water control," Petroleum Science and Technology, vol. 25, no. 3, pp. 353–371, 2007. · ·

19. O. Levenspiel, Chemical Reaction Engineering, John Wiley & Sons, New York, NY, USA, 3rd edition, 1999.

2

A New Approach to Hydraulic Stimulation of Geothermal Reservoirs by Roughness Induced Fracture Opening

Nima Gholizadeh Doonechaly[1], Sheik S. Rahman[1], and Andrei Kotousov[2]

[1]School of Petroleum Engineering, University of New South Wales, Sydney, Australia

[2]School of Mechanical Engineering, the University of Adelaide, South Australia, Australia

ABSTRACT

Hydraulic fracturing by shear slippage mechanism (mode II) has been studied in both laboratory and field scales to enhance permeability of geothermal reservoirs by numerous authors and their success stories have been reported. Shear slippage takes place along the planes of pre-existing fractures which causes opening of the fracture planes by the fracture asperities (roughness induced opening). Simplified empirical relationships, which are derived based on simple fracture experiments or best guess, are used to calculate compressive normal surface traction, residual aperture and shear displacement. This introduces ambiguity into the simulation results and often leads to erroneous predictions of reservoir performance.

In this study an innovative analytical approach based on the distributed dislocation technique is developed to simulate the roughness induced opening of fractures in the presence of compressive and shear stresses as well as fluid pressure inside the fracture. This provides fundamental basis for computation of aperture distribution for all parts of the fracture which can then be used in the next step of modeling fluid flow inside the fracture as a function of time. It also allows formulation of change in aperture due to thermal stresses. The stress distribution and the fluid pressure are calculated using the fluid flow modeling inside the fracture in a numerical framework in which thermo-hydro mechanical effects are also considered using finite element methods (FEM). In this study, fractures with their characteristic properties are considered to simulate rock deformation.

This new approach is applied to the Soultz-Sous-Forets geothermal reservoir to study changes in permeability and its impact on temperature drawdown. It has been shown that the analytical approach provides a more realistic prediction of residual fracture aperture which agrees well with the experience of existing EGS trials around the world. An average increase in aperture due to fluid induced shear dilation has been found to be lower and time required to obtain a sizeable reservoir volume is greater than those previously estimated.

INTRODUCTION

Reservoir Stimulation by Induced Fluid Pressure

Fractured reservoirs in crystalline rocks are usually stimulated by injected fluid pressure. As the injection of fluid continues the pressure inside the fractures increases gradually. The effective stress due to fluid pressure is expressed as:

$$\sigma_{eff} = \sigma_t - p$$

(1)

where σ_{eff} is the effective stress, σ_t is the total stress and p is the pore pressure. With further injection of fluid the effective shear stress, which is a function of effective stress, continuously decreases until it reaches a threshold value at which time it can no longer resist shear displacement of the fracture surfaces. At this stage the shear dilation will occur. During shear displacement rock fails by the shearing (Mode II) instead of opening (Mode I). In Mode II opening, the surface asperities of the rock slide over each other which cause more separation of the fracture surfaces. Such an interlocking of asperities increases the permeability of the rock. Any further increase in pressure can cause the effective closure stress to decrease to zero at which time the separation and interlocking of the fracture surfaces perpendicular to the fracture walls occur. The amount of pressure required to reach zero effective stress is highly dependent on the rock and fracture properties [1]. If the injection continues at some point it will exceed the tensile strength of the rock, which leads to tensile failure of the rock. This means that a certain level of permeability enhancement by shear displacement can be obtained. Mechanical representation of the shear displacement and the normal separation of the fracture surfaces can be described based on a specific failure criterion, such as Mohr-Coulomb (see Fig. 1). As the pressure inside the fracture increases the effective stress decreases: Mohr's circle moves towards the origin. As shown in Fig.1, when the minimum principal stress (closure stress) reaches zero the normal separation of fracture surfaces (Mode I) occurs. However, the shear dilation happens much earlier: when the Mohr's circle encounters the failure envelope (CD) at E. Shear dilation by induced fluid pressure was first detected in the laboratory experiments in 1970s. One of the earliest attempts by [2] showed a significant permeability increase by shear displacement. This observation was confirmed by [3] and [4]. Since then, shear dilation has been comprehensively studied in geotechnical and mining engineering. However, investigation of permeability enhancement by shear dilation in petroleum reservoirs began much later [5]. Since the shear dilation is caused by slippage of the asperities on top of each other, there is maximum dilation that can be reached. The maximum displacement that can be achieved is called characteristic height of the fracture [6]. Based on an experimental study the characteristic height is measured to be of the order of a fraction of a millimeter [7]. Fracture aperture that can be created by conventional hydraulic fracturing is in the order of tens of millimeters [8]. Reservoir rocks with rough surfaces and high shear strength are highly desirable for stimulation by shear displacement to work. One of the most comprehensive attempts to characterize the shear dilation caused by the fracture surface asperities was developed by [9]. In their model, the rock behavior was studied by considering fracture surface and its aperture, normal and shears closure and shear dilation. In another attempt, [10] proposed a methodology to obtain the mechanical aperture of the fractures. The authors used the methodology proposed by [11] to measure the aperture by a tapered feeler gauge using plane sawn surfaces to gain access to the joints. Mechanical aperture can be calculated using an empirical equation as proposed by [10]. Later [10] used the empirical equation proposed by [11] to model the normal closure of fracture surfaces based on the normal stress. [12] proposed an approach to describe the hydraulic and mechanical properties of the fracture including the shear dilation by induced fluid pressure.

$$\sigma_{eff} = \sigma_t - p$$

Figure 1: Mohr diagram describing the initiation of the shear dilation and normal fracture surface separation.

Mechanical models for shear displacement include displacement estimation under different stress boundary conditions in which a proper topographical model is used to describe the fracture surface. Also [13] experimentally studied the effect of normal stress and shear dilation on fluid flow properties of a naturally fractured core sample. They have used a servo-controlled axial/torsion load frame to test the fluid flow and mechanical behavior of the fracture surface during normal stress, slip and shear dilation. In another approach [14] proposed a semi-empirical correlation to determine the change in fracture aperture based on the amount of shear displacement between the fracture surfaces and the stress boundary condition. Also [15] extended the previous attempt of [14] by considering the effect of fracture propagation in shear dilation. in another attempt [16] used a linear relationship between shear displacement and the dilation of the fracture surfaces.

In this study, an analytical computational methodology based on distributed dislocation technique proposed by [6] is used to estimate the aperture distribution caused by the shear dilation in a fracture subject to different varying stress boundary conditions [6].

Two major assumptions are used in this approach to characterize the shear displacement of the fracture surfaces. The shear slippage between the fracture surfaces is described by using Coulomb friction law which explains the friction stress during the shear slippage based on the normal stress exerted on the fracture planes with a proportionality contact named friction factor as shown in Eq.

$$\tau = c + f\sigma_n$$

$$(2)$$

Where, τ_0 is the threshold shear stress value to initiate the shear slippage between the fracture surfaces. Also the friction factor, f, is dependent on the material properties, fracture geometry and surface asperities of the fracture [6]. Because a minor change in the fracture aperture causes a significant alteration of the fracture permeability estimation of the shear slippage of the fracture surfaces is of crucial importance in fluid flow simulation. In this study the coupling between the shear displacement and the change in fracture aperture is described by a step function. Fracture displacement normal to the fracture plane is simulated by using virtual springs distributed along the fracture length. Such springs are characterized by a specific spring constant which can be calculated numerically, experimentally or analytically [6]. Also the spring deformations are modeled in an elastic framework which results in the following system of equations describing the stress between the fracture surfaces:

$$\sigma_n = kE(\Delta - \delta_y) \text{ for } \delta_y < \Delta \qquad (3)$$

$$\sigma_n = 0 \text{ for } \delta_y > \Delta \qquad (4)$$

Where, Δ is the characteristic height of the fracture as shown in Eq. (3) and k is the spring constant. Equation (3) is associated with the rock compressibility and gives us the normal stress exerted on the fracture surfaces. After calculating the normal stresses on the fracture surfaces, the normal displacement of the surfaces is calculated by the distribution dislocation concept. Also the methodology proposed by [17] is used to calculate the spring constant based on a bed of nails as [17]:

$$k = E \frac{b\Delta}{L} \qquad (5)$$

Where, E id the Young modulus of elasticity, L is the fracture length and b is a constant less than unity. Also Eq. (4) implies the complete separation of the fracture surfaces in which no contact exists between the fracture asperities.

The complete set of boundary conditions for a fracture as shown in Fig. 2 are listed below [6]:

$$\sigma_n = kE(\Delta - \delta_y) \text{ for } \delta_y < \Delta$$

Figure 2: A schematic representation of a fracture subject to in-situ stress boundary conditions.

$$\sigma'_y = \sigma^\infty \text{ for } x^2 + y^2 \to \infty$$

$$(6)$$

$$\sigma'_y = kE(\Delta - \delta_y) + p \text{ for } c < |x| \le a$$

$$(7)$$

$$\sigma`y = p \text{ for } |x| \le c$$

$$(8)$$

$$\tau_n = \tau_0 + f \cdot kE(\Delta - \delta_y) \text{ for } c < |x| \leq a \tag{9}$$

$$\tau_n = 0 \text{ for } |x| \leq c \tag{10}$$

$$u_x = 0 \text{ for } |x| \geq a \tag{11}$$

$$u_y = 0 \text{ for } |x| \geq a \tag{12}$$

where σ'_y is the effective normal stress exerted on the fracture surfaces, k is the spring constant, p is the pore pressure, Δ is the characteristic height of the fracture, δ_y is the displacement of the fracture surfaces, E is the Modulus of elasticity, τ_n is the shear stress exerted on the fracture surfaces, τ_0 is the threshold stress requires to start the shear displacement of the fracture surfaces and u is the displacement. *As* mentioned above, the aperture distribution along the fracture surface is calculated based on an analytical methodology in which fracture geometry, stress distribution and fluid pressure inside the fracture are needed to be known as a priori. For this purpose a thermo-poro-elastic model is developed to simulate the fluid flow in the reservoir scale.

SIMULATION OF FLUID FLOW AND HEAT TRANSFER

Three distinct approaches exist in the literature to simulate the fluid flow in naturally fractured reservoirs namely: single continuum, dual continuum and discrete fracture approach. In single continuum, the fractured medium is represented by an equivalent homogeneous system using a specific permeability tensor. In dual continuum approach the whole domain is divided into two interacting domains: fractures and matrix where by matrix (represented by sugar cubes) provides the storage and fractures (having regular pattern) the permeability. In discrete fracture approach, fractures are explicitly discretized in the domain. These approaches are briefly discussed below followed by the proposed methodology which is used in this study.

Hybrid of Single Continuum and Discrete Fracture

Different approaches have been used in the literature to incorporate the fractures into the flow modeling. Each of these techniques has its own drawbacks and benefits. In this study a hybrid methodology combining the single continuum and discrete fracture networks model is used to increase accuracy and efficiency of the fluid flow simulation. In the proposed methodology a threshold value is defined for the fracture length. Fractures which are smaller than the threshold value are used to generate the grid based permeability tensor using boundary element technique. Fluid flow simulation is carried out by using the single continuum approach in the nominated blocks. Fractures which are equal to and longer than the threshold value are explicitly discretized in the domain using appropriate elements and the fluid flow is modeled using the discrete fracture approach. Such an approach provides a more accurate and realistic framework to consider the effect of long fractures on the fluid flow in fractured medium.

Domain Discretization Using the Hybrid Methodology

In this study the medium and long fractures ($l \geq$ 50m) are discretized using triangular elements and the contribution of flow by fractures ($l <$ 50m) are taken into account by calculating permeability tensor for each discretized element. A schematic representation of the domain discretization for a fractured reservoir is shown in Fig 3 (a) and (b).

Permeability tensor for each block is expressed as:

$$K = \begin{bmatrix} k_{xx} & k_{xy} \\ k_{yx} & k_{yy} \end{bmatrix}$$

$$(14)$$

Permeability tensors are calculated by simulating fluid flow in individual fractures in each element. The concept of permeability tensor was first introduced by [18] by considering a set of parallel fractures in a Representative Elementary Volume (REV) with zero matrix permeability [18]. In another attempt [19] developed a methodology for calculation of permeability tensor for arbitrary oriented fractures using superposition technique [19].

$$\sigma_n = 0 \, \text{for} \, \delta_y > \Delta$$

Figure 3: Domain discretization by using the hybrid of the single continuum and discrete fracture approach. (a) fractures equal to and longer than 50 m are explicitly discretized in the reservoir domain by using the triangular elements. (b) after the discretization of the long fractures, the effect of short fractures (<50m) are taken into account by calculation of the permeability tensor of the corresponding blocks which are cut by the fractures.

In this study the authors have considered interconnected fractures with fracture surface as infinite plate without roughness. In another approach [20] estimated permeability tensor by assuming fractures as a planar sink/source term [20]. Also [21] extended the approach and studied the effect of vertical fracture/ matrix permeability ratio on the permeability tensor. In a separate study, [22] used a numerical technique (BEM) to calculate the permeability tensor of the REV containing medium sized fractures considering fractures as a sink/source term [22]. Following this work [23] presented an analytical model to calculate the permeability tensor of the blocks containing infinite parallel fracture sets [23]. Also [24] improved the efficiency of their previous approach by considering the effect of short fractures using the analytical method proposed by [24]. In another approach [25] presented the first comprehensive methodology to calculate the permeability tensor for arbitrary oriented fractures in different length scales. In this study permeability tensor was determined by discretizing the solution domain into different subdomains depending on the length of the fractures using BEM [25]. Short fractures are considered as part of matrix porosity to improve the matrix permeability inhomogeneity. However, medium and long fractures are discretized explicitly in the domain and fluid flow is simulated using BEM. Then [26] extended [25] by increasing the efficiency of the BEM so that fluid flow in greater number of fractures can be simulated. The authors also presented for the first time effective permeability tensor calculation for the fractured REV by using the BEM. The effective permeability model was validated using laboratory derived data.

Rev Discretization for Permeability Tensor Calculation

To calculate the effective permeability tensor, the fractured REV is divided into three distinct regions: matrix (region 1), fracture (region 2) and region around the fractures (region 3) as shown in Fig. 4.

$$\sigma`y = p \ for|x| \le c$$

Figure 4: Domain discretization based on different fracture lengths.

Flow inside the fractures (region 2) is modeled using the cubic law. With the assumption of smooth fracture surfaces, cubic law can accurately simulate the flow inside the fractures [19, 27]. In matrix regions close to the fractures (region 3), the Darcy equation Eq. (14) is coupled with the mass conservation equation to consider the effect of the fracture on the flow of fluid in the region close to the fractures. Size of this region depends on the size of the fracture. Also fluid flow simulation in the matrix (region 1) is described as follow:

$$V_f = -K_f \nabla P_f$$

(14)

$$\frac{\partial}{\partial L}(k_f \frac{\partial p}{\partial L}) + Q_f + q_{ff} = 0$$

(15)

where p_f is the fluid pressure inside the fractures and p is the pore fluid pressure. For the short fractures which are considered as part of the matrix porosity, the Laplace equation is solved using the following boundary conditions:

$$p_{mi} = p_{fi}$$

(16)

$$v_{mi} = v_{fi}$$

(17)

Where, p_{mi} is the matrix pressure and p_{fi} is the fracture pressure at the matrix/fracture interface and v_{mi} is the normal fluid velocity at the ith fracture node along the fracture surface. Since the pressure on the matrix fracture interface is unknown, periodic boundary condition is applied in an iterative scheme to calculate the pressure values.

Reservoir Scale Fluid Flow Simulation

Fluid flow in long fractures (l>50m) is coupled with discretized element based permeability tensor in poro-thermo-elastic environment by using local-thermal non-equilibrium.

Different numerical techniques have been used to model thermo-poro-elastic phenomena in fractured porous media. To have a detailed understanding of the complex geo mechanical aspects of the fractured rocks and the induced perturbation, such as thermal drawdown caused by the cold injection fluid in geothermal reservoirs an appropriate numerical technique should be used which is capable of (a) adequately applying the boundary and initial conditions and (b) accurately representing the system geometry. In order to take the aforementioned issues into account, FEM is used in the current study.

Weighted residual method and the Green's theorem are applied to discretize the mass, momentum and energy conservations equations [28]. As mentioned before, the finite element method is used in this study for the numerical simulation purpose. Therefore the state variables namely: displacement, pore pressure and temperature are defined using proper shape functions as:

$$u = N_u \bar{u}$$

(18)

$$p = N_p \bar{p}$$

(19)

$$T = N_T \bar{T}$$

(20)

Where N is the corresponding shape function and $\bar{u}$, $\bar{p}$ and $\bar{t}$ are the nodal values of the corresponding state variable. By applying the Galerkin's method and replacing the weighting functions by the corresponding variables' shape functions, the discretized form of the conservation equations can be written as follow [29, 30]:

$$\left(K + \frac{G}{3}\right)\nabla(\nabla \cdot u) + G \cdot \nabla^2 u - \alpha \nabla p - \gamma_1 \nabla T_m = 0$$

(21)

$$\alpha(\nabla \cdot \dot{u}) + \beta \dot{p} - \frac{k}{\mu}\nabla^2 p - \gamma_2 \dot{T} = 0$$

(22)

$$\dot{T} + v(\nabla T) - c^T \nabla^2 T = 0$$

(23)

Where, K is the bulk modulus of elasticity, G is the shear modulus, γ_1 and γ_2 are the thermal expansion coefficient of the fluid and solid respectively; k is the permeability T_m is the matrix temperature, T is the fluid temperature and μ is the fluid viscosity.

FRACTURE NETWORK GENERATION

Simulation of naturally fractured reservoirs offers significant challenges due to the lack of a methodology that can utilize field data. To date several methods have been proposed in the literature to characterize naturally fractured reservoirs. In this study a hybrid tectono-stochastic simulation is proposed to characterize a naturally fractured reservoir [31]. A finite element based model is used to simulate the tectonic event of folding and unfolding of a geological structure. A nested neuro-stochastic technique is used to develop the inter-relationship between different sources of data (seismic attributes, borehole images, core description, well logs etc.) and at the same time the sequential Gaussian approach is utilised to analyze field data along with fracture probability data. This approach has the ability to overcome commonly experienced discontinuity of the data in both horizontal and vertical directions.

RESULTS AND DISCUSSIONS

The proposed methodology is used to generate the discrete fracture map of the Soultz geothermal reservoir at the depth of 3650 m. the statistical parameters used to generate the discrete fracture map is shown in Table 1.

Table 1: Statistical data used for the discrete fracture network generation After [32]

Fracture set	Azimuth			Dip				Fracture No.	Radius (m)	Transmissivity (m2\s)
	Distribution Law	Mean	Half-Width	Distribu-tion Law	Mean	Half-Width	Dip Direction			
F1	Normal	2	16	normal	70	7	NW	1.3E-7	187	6E-6
F2	Normal	162	19	normal	70	7	NE	3E-9	150	6E-6
F3	Normal	42	6	normal	74	3	NW	1.76E-8	95	4E-6
F4	Normal	129	6	normal	68	3	SW	3.3E-8	112	2E-6
F5	Uniform	0	180	normal	70	9	-	1E-8	100	5E-7

The discrete fracture map, the corresponding mesh generated for the reservoir domain and the permeability tensors for each triangular element (a sample region which is cut by a fracture of length<50m) are shown in Fig. 5 (a), (b) and (c) respectively.

Also the reservoir properties used for the stimulation purpose are shown in Table 2. The reservoir is pressurized by injecting fluid through the injection well (GPK2). The pressurization was carried out over a period of 52 weeks. During the pressurization, the change in fracture width for each individual natural fracture and the resulting permeability tensor were calculated. Following stimulation of the reservoir, a flow test was carried out over a period of 14 years. During the flow test, changes in fracture apertures due to thermo-poro-elastic stresses and the consequent changes in permeability were determined. Also estimated were the thermal drawdown, produced fluid temperature and production rate of the Soultz EGS.

$$\tau_n = \tau_0 + f \cdot kE(\Delta - \delta_y) \text{ for } c < |x| \leq a$$

Figure 5. a) discrete fracture network at the depth of 3650 m (b) the corresponding discretization for the fractures longer than 50 m and (c) permeability tensor for a sample fracture (<50m).

Results of shear dilation are presented as average percentage increase in fracture aperture (see Fig. 6). From Fig. 6, it can be seen that there exists three distinct aperture histories: 0-40 weeks, 40-50 weeks and 50 weeks and above. Until about 40 weeks, a slow but linear increase in occurrence of dilation events due to induced fluid pressure of 51.7 MPa (bottom hole) and reaches a value of about 18% (average increase in aperture). Following this time, the rate of occurrence of dilation events increases sharply until about 50 weeks, thus reaching 60% increase in average fracture aperture. After which, no significant dilation events can be observed (a plateau of events is reached). When compared with previous study [29], in which shear dilation events are estimated based on a semi-empirical model (Willis-Richards et al, 1996), it can be seen that the time required to overcome the threshold stress is 40 weeks which is about 12 weeks longer than the previous studies. Also the time requires for an increase in the average fracture aperture of 58 % is about 8 weeks longer than that predicted by the previous study. During the flow test, changes in fracture apertures due to thermo-poro-elastic stresses and the consequent changes in permeability were determined. Also estimated were the thermal drawdown, produced fluid temperature and production rate of the Soultz EGS.

Table 2: Stress and reservoir data for strike-slip stress regime at Soutlz geothermal reservoir

Rock Properties	
Young's modulus (GPa)	40
Poisson's ratio	0.25
Density (kg/m3)	2700
Fracture basic friction angle (deg)	40
Shear dilation angle (Deg)	2.8
90% closure stress (MPa)	20
In situ mean permeability (m2)	9.0×10^{-17}
Fracture properties	
Fractal Dimension, D	1.2
Fracture density (m2/m3)	0.12
Smallest fracture radius (m)	15
Largest fracture radius (m)	250
Fracture Permeability	0.3×10^{-15}
Stress data	
Maximum horizontal stress (MPa)	78.9
Minimum horizontal stress (MPa)	53.3
Fluid properties	
Density (kg/m3)	1000

Viscosity (Pa s)	3 x 10-4
Hydrostatic fluid pressure (MPa)	34.5
Injector pressure, stimulation (MPa)	51.7
Injector pressure, production (MPa)	44.8
Producer pressure, stimulation (MPa)	N/A
Producer pressure, production (MPa)	31.0
Other reservoir data	
Well radius (m)	0.1
Number of injection wells	1
Number of production wells	2
Reservoir depth (m)	3650

$$\tau_n = 0 \; \text{for} \; |x| \leq c$$

Figure 6: Comparison of Average aperture increase between the current approach and the previous study.

The locations of the dilation events during the stimulation period are shown in Fig. 7. As shown in this figure, after 40 weeks of stimulation about half of the reservoir is affected by the shear dilation and after 52 weeks of injection shear dilation happened in almost all parts of the reservoir.

Also the reservoir pressure and stress distribution profiles (see Figs. 8 and 9) show that after 40 weeks of stimulation the injected fluid pressure affected almost all of the fractures and that after 52 weeks of injection the pressure is established in all part of the reservoir domain. Similarly the x- and y component of the effective stress decreased significantly over the entire reservoir domain towards the end of the stimulation period.

$$\frac{\partial}{\partial L}\left(k_f \frac{\partial p}{\partial L}\right) + Q_f + q_{ff} = 0$$

Figure 7: Location of the dilation events marked by the dots after (a) 1 week (b) 40 weeks and (c) 52 weeks of stimulation with σH = 78.9 MPa and σh = 53.3 MPa, Pinj = 51.7 MPa.

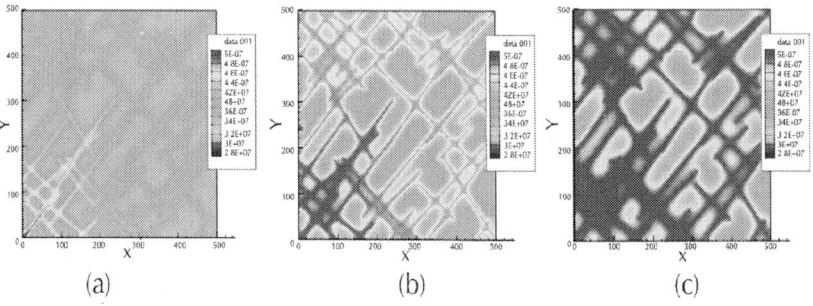

(a)   (b)   (c)

Figure 8: Pore pressure distribution of the fractured reservoir at different stimulation stages: after (a) 1 week, (b) 40 weeks and (c) 52 weeks for a strike slip stress regime with σH = 78.9 MPa and σh = 53.3 MPa, Pinj = 51.7 MPa.

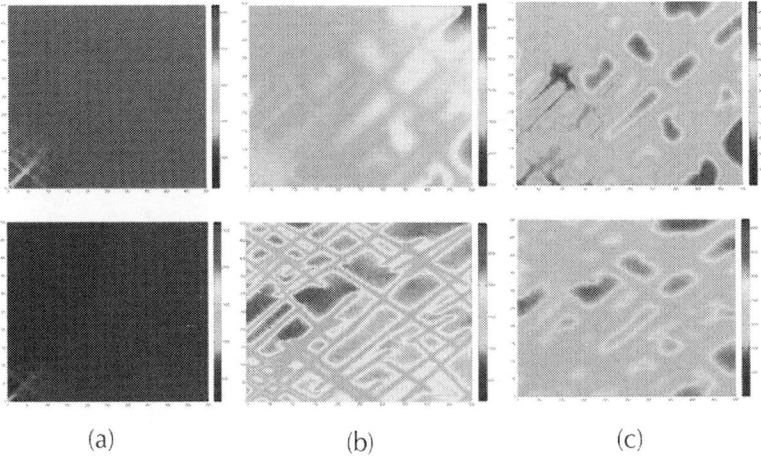

(a)　　　　　　　　(b)　　　　　　　　(c)

Figure 9: x (top) and y (bottom) components of effective stress after: (a) 1 week, (b) 40 weeks and (c) 52 weeks of stimulation for σH = 78. 9 MPa and σh = 53.3 MPa, P_{inj} = 51.7 MPa.

After the stimulation period a numerical experiment is carried out to assess the produced matrix temperature for 14 years of cold fluid circulation. Because of the low fluid and rock matrix contact area at the early stage of production, the heat transfer and the resulting thermal drawdown is very low (see Fig 10 a). With the pass of time the fluid sweeps over a large part of the reservoir which increases thermal drawdown. At the end of the 14 years of production the average matrix temperature drops from 200 to 150°C which is quite low (drop of 50°C) compared to previous studies (drop of 80°C over the production period of 14 years as in [29]) under the same reservoir conditions. Also in Fig. 10(bottom) the Log10 RMS fluid velocity profile after 1 year, 10 years and 14 years of production are presented. From the results it can be observed that during the early production period (1 year) high pore pressure is primarily built up around the injection well and the flow of fluid is primarily through major inter-connected flow paths. With the progress of time the injection pressure advances towards the production well. After 14 years of production, the fluid sweeps through a significant part of the reservoir. Also the x- and y components of effective stress distribution of the Soultz geothermal reservoir during different stages of production are shown in Fig. 11. These results show that by the end of 14 years of production the effective stresses throughout the reservoir are significantly reduced, thus allowing most fractures to open and conduct fluid. The reduction in the effective stresses is caused by the cold circulating fluid as well as thermal drawdown.

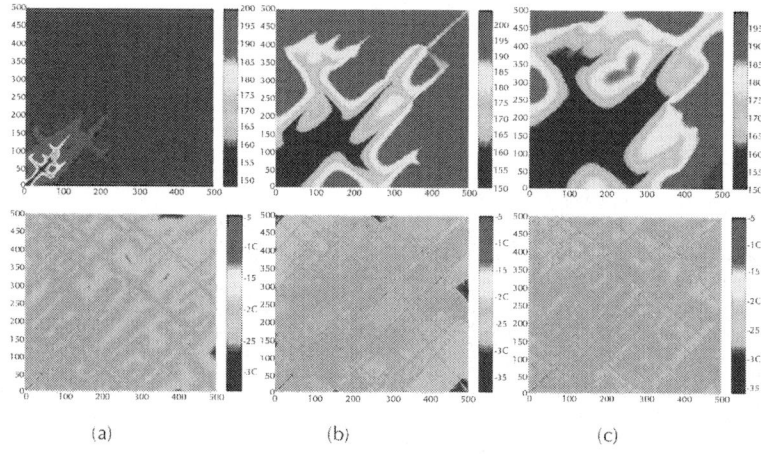

(a)   (b)   (c)

Figure 10: Reservoir temperature profile (top) and Log10RMS fluid velocity profile (bottom) after (a) 1 year (b) 10 years and (c) 14 years of production with $\sigma H = 78.9$ MPa and $\sigma h = 53.3$ MPa, $P_{inj}=44.8$ MPa and $P_{prod}=31$ MPa.

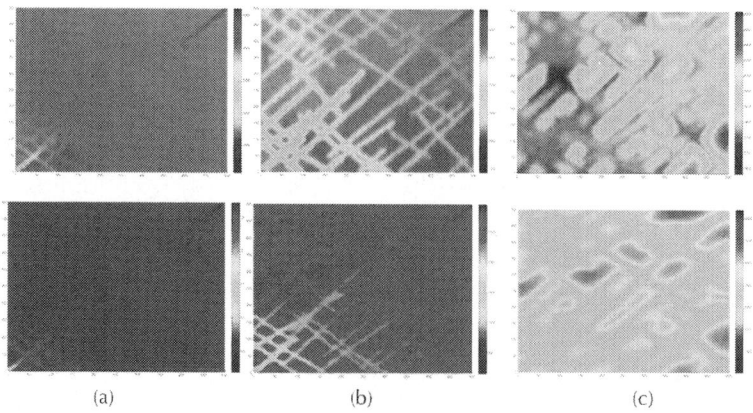

(a)   (b)   (c)

Figure 11: x (top) and y (bottom) component of effective stress after (a) 1 year (b) 10 years and (c) 14 years of production with $\sigma H = 78.9$ MPa and $\sigma h = 53.3$ MPa, $P_{inj}=44.8$ MPa and $P_{prod}=31$ MPa.

CONCLUSIONS

In this paper, a roughness induced shear displacement model in a poro-thermoelastic environment combined with an advanced computational technique is used to study the effects of induced fluid pressure and thermal stresses (cooling effect) on reservoir permeability and consequent increase in hot water production. It has been shown that surface roughness induced shear displacement provides a more realistic prediction of residual fracture aperture. These results agree well with the experience of existing EGS trials around the world. An average increase in aperture due to fluid induced shear dilation has been found to be lower and time required to obtain a maximum stimulated volume is greater. Results of this study are in consistent with that of previous studies: for every geothermal system there exists an optimum injection schedule (injection pressure and duration). Any further increases in stimulation effort, i.e. stimulation time for a given stimulation pressure, does not provide additional permeability enhancement.

REFERENCES

1. Roshan, H. and S.S. Rahman, Effects of Ion Advection and Thermal Convection on Pore Pressure Changes in High Permeable Chemically Active Shale Formations. Petroleum Science and Technology, 2013. 31(7): p. 727-737.

2. Lockner, D.A., J.B. Walsh, and J.D. Byerlee, Changes in seismic velocity and attenuation during deformation of granite. Journal of Geophysical Research, 1977. 82(33): p. 5374-5378.

3. Hast, N., Limits of stress measurements in the Earth's crust. Rock mechanics, 1979. 11(3): p. 143-150.

4. Solberg, P., D. Lockner, and J.D. Byerlee, Hydraulic fracturing in granite under geothermal conditions. International Journal of Rock Mechanics and Mining Sciences & Geomechanics Abstracts, 1980. 17(1): p. 25-33.

5. Rahman, M.K., M.M. Hossain, and S.S. Rahman, An analytical method for mixed-mode propagation of pressurized fractures in remotely compressed rocks. International Journal of Fracture, 2000. 103(3): p. 243-258.

6.   Kotousov, A., L. Bortolan Neto, and S. Rahman, Theoretical model for roughness induced opening of cracks subjected to compression and shear loading. International Journal of Fracture, 2011. 172(1): p. 9-18.

7.   Heidinger, P., J. Dornstädter, and A. Fabritius, HDR economic modelling: HDRec software. Geothermics, 2006. 35(5–6): p. 683-710.

8.   Blumenthal, M., et al., Hydraulic model of the deep reservoir quantifying the multi-well tracer test.. EHDRA Scientific Conference, Soultz-sous-Forets, 2007.

9.   Barton, N. and V. Choubey, The shear strength of rock joints in theory and practice. Rock mechanics, 1977. 10(1-2): p. 1-54.

10.   Barton, N., S. Bandis, and K. Bakhtar, Strength, deformation and conductivity coupling of rock joints. International Journal of Rock Mechanics and Mining Sciences & Geomechanics Abstracts, 1985. 22(3): p. 121-140.

11.   Bandis, S., Experimental studies of scale effects on shear strength and deformation of rock joints. PhD Thesis, 1980.

12.   Piggott, A.R. and D. Elsworth, A Hydromechanical Representation of Rock Fractures, 1991, A.A. Balkema. Permission to Distribute - American Rock Mechanics Association.

13.   Olsson, W.A. and S.R. Brown, Hydromechanical response of a fracture undergoing compression and shear. International Journal of Rock Mechanics and Mining Sciences & Geomechanics Abstracts, 1993. 30(7): p. 845-851.

14.   Willis-Richards, J., K. Watanabe, and H. Takahashi, Progress toward a stochastic rock mechanics model of engineered geothermal systems. J. Geophys. Res., 1996. 101(B8): p. 17481-17496.

15.   Rahman, M.K., M.M. Hossain, and S.S. Rahman, A shear-dilation-based model for evaluation of hydraulically stimulated naturally fractured reservoirs. International Journal for Numerical and Analytical Methods in Geomechanics, 2002. 26(5): p. 469-497.

16.   Zhang, X., R.G. Jeffrey, and E. Detournay, Propagation of a hydraulic fracture parallel to a free surface. International Journal for Numerical and Analytical Methods in Geomechanics, 2005. 29(13): p. 1317-1340.

17. Gangi, A.F., Variation of whole and fractured porous rock permeability with confining pressure. International Journal of Rock Mechanics and Mining Sciences & Geomechanics Abstracts, 1978. 15(5): p. 249-257.

18. Snow, D.T., Anisotropie Permeability of Fractured Media. Water Resources Research, 1969. 5(6): p. 1273-1289.

19. Long, J.C.S., et al., Porous media equivalents for networks of discontinuous fractures. Water Resources Research, 1982. 18(3): p. 645-658.

20. Baumgartner, J., P.L. Moore, and A. Gtrard, Drilling of Hot and Fractured Granite at Soultz-sous-Forgts (France). Proceedings of the World Geothermal Congress, Florence, Italy,International Geothermal Association,, 1995. 4: p. 2657-2663.

21. Rasmussen, T.C., J. Yeh, and D. Evans, Effect of variable fracture permeability/matrix permeability ratios on three-dimensional fractured rock hydraulic conductivity. Proceedings of the Conference on Geostatistical, Sensitivity, and Uncertainty Methods for Ground-Water Flow and Radionuclide Transport Modeling, San Francisco, California, September 1987, B. E. Buxton, Batelle Press, Columbus, OH, 1987, 337, 1987.

22. Lough, M.F., S.H. Lee, and J. Kamath, A New Method To Calculate Effective Permeability of Gridblocks Used in the Simulation of Naturally Fractured Reservoirs. SPE Reservoir Engineering, 1997. 12(3): p. 219-224.

23. Chen, M., M. Bai, and J.C. Roegiers, Permeability Tensors of Anisotropic Fracture Networks. Mathematical Geology, 1999. 31(4): p. 335-373.

24. Lee, S.H., M.F. Lough, and C.L. Jensen, Hierarchical modeling of flow in naturally fractured formations with multiple length scales. Water Resources Research, 2001. 37(3): p. 443-455.

25. Teimoori, A., et al., Effective Permeability Calculation Using Boundary Element Method in Naturally Fractured Reservoirs. Petroleum Science and Technology, 2005. 23(5-6): p. 693-709.

26. Fahad, M., S.S. Rahman, and Y. Cinar, A Numerical and Experimental Procedure to Estimate Grid Based Effective Permeability Tensor for Geothermal Reservoirs. Geothermal Resources Council Transactions, 2011.

27. Rasmussen, M.L. and F. Civan, Full, Short-, and Long-Time Analytical Solutions for Hindered Matrix-Fracture Transfer Models of Naturally Fractured Petroleum Reservoirs, in SPE Production and Operations Symposium2003, Society of Petroleum Engineers: Oklahoma City, Oklahoma.

28. Bathe, K.J., Finite element procedures. 1996: Prentice Hall.

29. Koh, J., H. Roshan, and S.S. Rahman, A numerical study on the long term thermo-poroelastic effects of cold water injection into naturally fractured geothermal reservoirs. Computers and Geotechnics, 2011. 38(5): p. 669-682.

30. Gholizadeh Doonechaly, N., S.S. Rahman, and A. Kotousov, An Innovative Stimulation Technology for Permeability Enhancement in Enhanced Geothermal System--Fully Coupled Thermo-Poroelastic Numerical Approach. 36th Geothermal Resources Council Transactions, 2012.

31. Gholizadeh Doonechaly, N. and S.S. Rahman, 3D hybrid tectono-stochastic modeling of naturally fractured reservoir: Application of finite element method and stochastic simulation technique. Tectonophysics, 2012. 541–543(0): p. 43-56.

32. Genter, A., et al., Contribution of the exploration of deep crystalline fractured reservoir of Soultz to the knowledge of enhanced geothermal systems (EGS). Comptes Rendus Geoscience, 2010. 342(7–8): p. 502-516.

Chapter 3

Compositional Simulation on the Flow of Polymeric Solution Alternating CO$_2$ through Heavy Oil Reservoir

Moon Sik Jeong, Jinhyung Cho, Jinsuk Choi, Ji Ho Lee, and Kun Sang Lee

Department of Natural Resources and Environmental Engineering, Hanyang University, 222 Wangsimni-ro, Seongdong-gu, Seoul 133-791, Republic of Korea

ABSTRACT

Water-alternating-gas (WAG) method provides superior mobility control of CO_2 and improves sweep efficiency. However, WAG process has some problems in highly viscous oil reservoir such as gravity overriding and poor mobility ratio. To examine the applicability of carbon dioxide to recover viscous oil from highly heterogeneous reservoirs, this study suggests polymer-alternating-gas (PAG) process. The process involves a combination of polymer flooding and CO_2 injection. In this numerical model, high viscosity of oil and high heterogeneity of reservoir are the main challenges. To confirm the effectiveness of PAG process in the model, four processes (waterflooding, continuous CO_2 injection, WAG process, and PAG process) are implemented and recovery factor, WOR, and GOR are compared. Simulation results show that PAG method would increase oil recovery over 45% compared with WAG process. The WAG ratio of 2 is found to be the optimum value for maximum oil recovery. The additional oil recovery of 3% through the 2 WAG ratios is achieved over the base case of 1:1 PAG ratio and 180 days cycle period.

INTRODUCTION

Recently, interest in CO_2 flooding has grown as a method of enhanced heavy oil recovery. Injected CO_2 can extract the heavy oil components by oil swelling and viscosity reduction. However, the mobility ratio of CO_2 is unfavorable to recover heavy oils. It causes viscosity fingering and gravity override through heterogeneous reservoirs. These phenomena make an early breakthrough of injected CO_2 and reduce oil recovery. The problems led by poor viscosity ratio are more severe in heavy oils than light oils. Although the CO_2 flooding has been applied and its success has been reported in many heavy oil cases [1–7], there still remain the aforementioned problems that need to be solved in order to implement the CO_2 injection in the heavy oil reservoirs.

Mobility control in CO_2 flooding is very important to solve the low recovery efficiency problem. CO_2 injection method can achieve higher microscopic displacement efficiency than those of other processes. However, viscosity of CO_2 is usually about 1/10 that of oil in the reservoir conditions [8]. As a result, the sweep efficiency of CO_2 flooding is lower than efficiency of waterflooding. The water-alternating-gas (WAG) process is suggested by Caudle and Dyes [9] to improve sweep efficiency of CO_2 injection. Alternating or coinjection of CO_2 and water enhances the recovery of oil. The injected water increases sweep efficiency and stabilizes the gas front. When slugs of CO_2 and water are injected into reservoir consecutively, some part of CO_2 is dissolved in the oil and reduces the oil viscosity. Thus, the mobility ratio between displacing and displaced fluid is decreased. It becomes favorable condition to control the CO_2 breakthrough and improve recovery efficiency.

Another suggested technique which advances sweep efficiency for the heterogeneous reservoir including high permeable thief zones is integrated polymer and CO_2 flooding. Generally, polymer flooding is known as effective process when mobility ratio of waterflooding is high, the heterogeneity of reservoir is high, or both of them exist [10]. Polymer flooding is processed by adding polymer into the water to decrease mobility of displacing fluid. Dissolved polymer increases the viscosity of displacing fluid and decreases the effective permeability of aqueous phase through adsorption. High adsorption of polymer through mainly high permeable streaks reduces permeability so that it induces diverting displacing fluid into low permeable zones and increases the oil recovery. However, polymer flooding is not a great way to decrease residual oil saturation. The polymer degradation and shear effect have been problems in application of polymer flood. A substantial amount of polymer is required to reduce the unsuitably high viscosity ratio to a value of approximately one in the heavy oil reservoirs. The significant required number of polymers in such reservoirs leads to high cost [11].

To overcome these problems, such as viscous fingering, poor sweep efficiency, and polymer concentration, integrated EOR method as coupling polymer flooding and CO_2 flooding is of importance. It has both advantages of CO_2 flooding and polymer flooding, solubility of CO_2 injection and mobility control of polymer injection. According to Zhang et al. [11], polymer/gas-alternating-water (PGAW) is combination of these two methods. Majidaie et al. [12] simulated chemically enhanced water-alternating-gas (CWAG) injection in homogeneous reservoir. Li et al. [13] carried out a case study of polymer-alternating-gas (PAG) simulation. However, more research for coupling CO_2 flooding and polymer injection is still needed. The previous simulation studies [12, 13] have been carried out in light oil reservoir. Although Zhang et al. [11] assessed its performance considering heavy oil; it is limited with experimental scale. Applications of PAG process in heavy and heterogeneous reservoirs have not been conducted sufficiently. For this reason, specific purpose of this study focused on the simulation of PAG process in field scale heterogeneous reservoir containing heavy oil. To evaluate the effectiveness of PAG process in the model, four processes (waterflooding, continuous CO_2injection, WAG process, and PAG process) are implemented and analyzed with oil recovery factor, WOR, and GOR. In addition, PAG ratio and PAG cycle have been parameterized to maximize the performance of PAG.

NUMERICAL SIMULATION

Fluid Modeling

The oil properties of Schrader Bluff and West Sak are referenced for viscous oil modeling. Composition of the oil is reported in Table 1. The portions of intermediate components are small and heavy components are main part. Properties and viscosity data which are used for regression analysis are based on the literature study of Ning et al. (Tables 2 and 3) [14]. Peng and Robinson [15] method is applied to generate PVT data of referenced components.

Table 1: Composition of viscous oil

Component	Mole fraction
CO_2	0.00027
N_2 to C_1	0.30446
C_2 to C_4	0.01018
C_5 to C_7	0.02464
C_8 to C_{12}	0.09672
C_{13} to C_{19}	0.21201
C_{20} to C_{30}	0.35172
Total	1

Table 2: Properties of reservoir fluid

Stock tank oil density	0.953 kg/m^3
STO API gravity	16.9
Gas oil ratio	32.2 m^3/m^3
Saturation pressure	101 atm

Table 3: Viscosity of the reservoir fluid at 24°C

Pressure (atm)	Viscosity (kg/ m·sec)
170	0.1411
136	0.1300
116	0.1250
109	0.1225

Due to solubility of CO_2 into heavy oil, K-values are calculated to represent an equilibrium state between components. The definition of K-value is the ratio of equilibrium gas component Y_i to the equilibrium liquid composition x_i as follows:

$$K_i \equiv \frac{y_i}{x_i}.$$

(1)

K_i is a function of pressure, temperature, and oil composition. K-values are calculated by satisfying the fugacity of equilibrium state based on EOS model. For the oil components given in Table 1, K-values are estimated on various pressures as depicted in Figure 1.

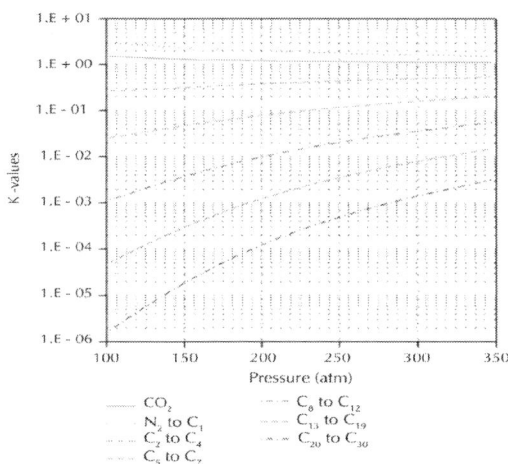

Figure 1: Estimations of -values for reservoir oil components at 24°C.

Hypothetical Reservoir Modeling

The hypothetical reservoir model is assumed as layered model which is discretized into 50 × 1 × 10 grid blocks. Each grid block has dimension of 1.2 m × 3 m × 1.5 m (Figure 2). The depth of reservoir is 244 m and reference pressure and temperature at this point are 121 atm and 24°C. The porosity is 30%. Average permeability is $5.3 \times 10^{-8} m^2$ which has Dykstra-Parsons coefficient (V_{DP}) representing variation of permeability as 0.75 which is determined by permeability variation [16] as follows:

$$V_{DP} = \frac{k_{50} - k_{84.1}}{k_{50}},$$

(2)

where K_{50} is permeability value at 50% probability and $K_{84.1}$ is permeability value at 84.1% of the cumulative sample. The range of coefficient varies from 0 to 1. If the heterogeneity of reservoir increases, the value of coefficient approaches to 1. Vertical/horizontal permeability ratio is assumed as 0.1. The initial water saturation is 0.2 and oil saturation is 0.8. Viscosity of water is 0.00045 Kg/m sec and oil and CO_2 viscosities are estimated to be about 0.094 Kg/m sec and 0.0001 Kg/m sec , respectively. The viscosity of polymeric solution is 0.022 Kg/m sec at the concentration of 1,000 ppm. Tables 4 and 5 present the input reservoir properties and permeability data used for this simulation. Water is injected during the first year and other processes (waterflooding, continuous CO_2 injection, WAG process, and PAG process) are implemented for next 10 years.

Table 4: Input data for reservoir simulation

Parameters	Values
Reservoir size (m³)	60 × 3 × 15
Number of grids	50 × 1 × 10
Permeability	
Average (m²)	5.3 × 10⁻⁸
K_v/K_h	0.1
V_{DP}	0.75

Porosity	0.3
Pressure (atm)	121
Temperature (°C)	24
Initial saturation	
Water	0.2
Oil	0.8
Viscosity (kg/ m·sec)	
Water	0.00045
Oil	0.094
CO_2	0.0001
Polymer 1,000 ppm	0.022

Table 5: Permeability data for layered reservoir

Layer number	Permeability $(10^{-7}\,m^2)$
1	2.4
2	1.6
3	2.6
4	1.3
5	0.89
6	0.69
7	0.44
8	0.20
9	0.15
10	0.08

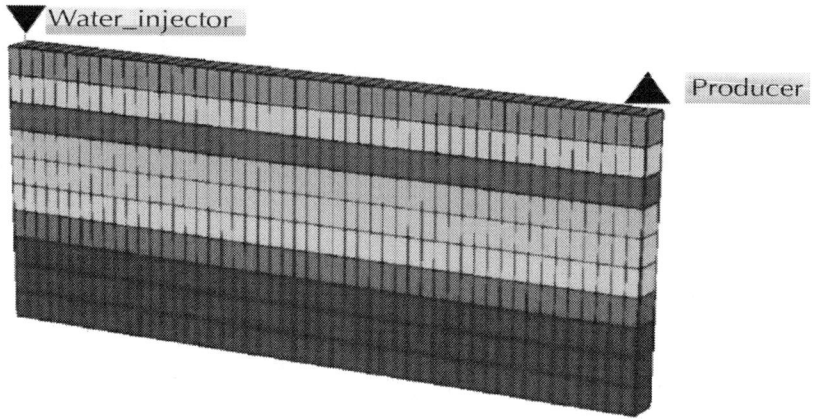

Figure 2: Hypothetical model consisting of different permeability layers.

Mobility Control

The objective of WAG process is originally to aim for the ideal oil recovery system: improvements of macroscopic and microscopic sweep efficiency at once. The injected water (or polymeric solution) is able to control the injected gas mobility as follows:

$$f_w = \frac{k_w/\mu_w}{k_w/\mu_w + k_o/\mu_o + k_g/\mu_g},$$

$$f_g = \frac{k_g/\mu_g}{k_w/\mu_w + k_o/\mu_o + k_g/\mu_g},$$

(3)

Where f is the fractional flow, K is the permeability, and μ is the viscosity [17].

The oil recovery factor (R_f) is determined by microscopic sweep efficiency and the macroscopic sweep efficiency. The macroscopic sweep efficiency can be described by the horizontal and vertical sweep efficiencies. The recovery factor is formulated by

$$R_f = E_v E_h,$$

(4)

Where E_v is the vertical sweep efficiency and E_h is the horizontal sweep efficiency [10].

The mobility ratio (5) [18] affects horizontal sweep efficiency and the vertical sweep efficiency is related to the ratio of viscous to gravity forces (6) [19]. Consider

$$M = \frac{k_{r,\text{displacing fluid}}/\mu_{\text{displacing fluid}}}{k_{r,\text{displaced fluid}}/\mu_{\text{displaced fluid}}},$$

(5)

$$R_{v/g} = \left(\frac{\nu\mu_o}{kg\Delta\rho}\right)\left(\frac{L}{h}\right),$$

(6)

where M is the mobility ratio, $R_{v/g}$ is the viscous/gravity forces ratio μ_o is Darcy velocity, is the oil viscosity, g is the permeability, is the gravitational acceleration, $\Delta\rho$ is difference in oil and solvent densities, L is distance between wells, and h is height of reservoir.

Polymer Behavior

The polymer adsorption at reservoir rock could be described by Langmuir-type isotherm [20] such as

$$ad = \frac{(a_1 + a_2 S_b) C_p}{1 + a_3 C_p},$$

(7)

Where a_1, a_2, and a_3 are coefficients of isothermal Langmuir equation, S_b is the salinity of the brine, and C_p is the mole fraction of polymer. Adsorption is assumed as irreversible process. By means of adsorption, not only more polymer concentration is required to reach target polymer concentration, but also induced reduction of permeability decreases flow capacity [21].

RESULTS AND DISCUSSION

Comparison of Processes

This study aims to confirm the effectiveness of PAG process in the heavy oil reservoirs. To examine the performance of various injection processes such as waterflooding, continuous CO_2 injection, WAG process, and PAG process, oil recovery factors are compared as depicted in Figure 3. Oil recovery from waterflooding is slightly higher than recovery of CO_2 flooding. CO_2 flooding has better recovery efficiency than that of waterflooding until the recovery factor reaches 19%. The efficiencies of CO_2 flooding and waterflooding are reversed after that point. The reservoir considered in this simulation includes high permeable layer at the top. Gravity overriding effect and early breakthrough mainly occur through the high permeable streak. Figure 4indicates the gravity overriding effect and CO_2 breakthrough after one year of CO_2 injection. The breakthrough can develop main CO_2 flow path and most of injected CO_2 passes through the path. Despite high potential for displacement efficiency, this effect reduces the sweep efficiency in application of CO_2flooding. As this phenomenon makes no more increases in oil recovery after five years of CO_2 injection, oil recoveries between waterflooding and CO_2 flooding are reversed.

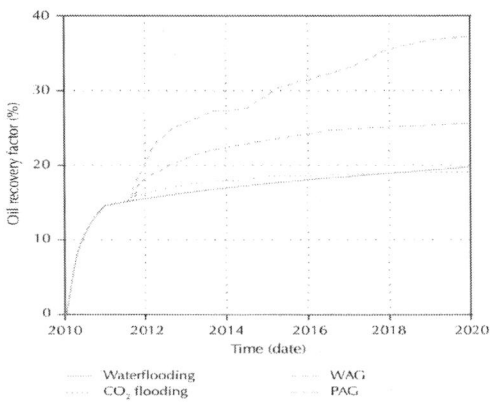

Figure 3: Oil recovery factors for different processes.

Water_injector   Producer

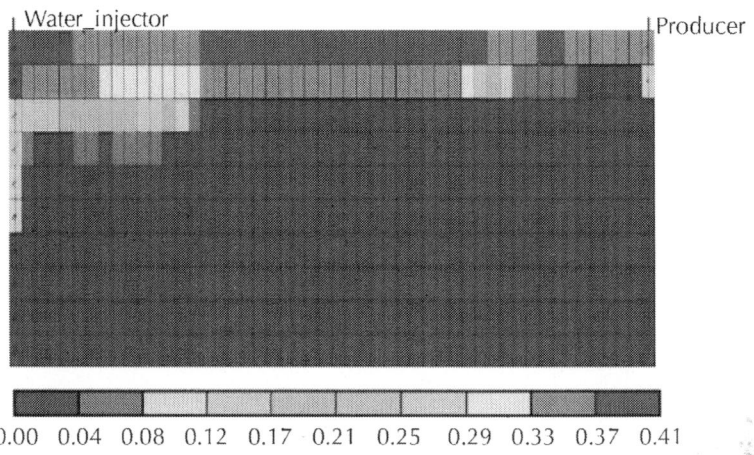

0.00   0.04   0.08   0.12   0.17   0.21   0.25   0.29   0.33   0.37   0.41

Figure 4: CO_2 mole fraction after one year of CO_2 injection in continuous CO_2 flooding.

WAG process is implemented and investigated. WAG ratio is set as 1 : 1 and one cycle period is 180 days, respectively. According to Figure 3, WAG process obtains 26% oil recovery while recovery factors of waterflooding and CO_2 flooding are less than 20%. This improved oil recovery as much as 6% by application of WAG is reasoned from the increased sweep efficiency and displacement efficiency by applying water and CO_2 flooding. A great amount of oil is easily extracted to the producer.

In PAG process, polymeric solution is injected into reservoir instead of water in WAG process. The solution contains 1,000 ppm polymer and it could prove the effect of polymeric solution during PAG process. PAG process achieves the highest oil recovery in Figure 3. PAG process takes 37% recovery factor. The enhancement of oil recovery by PAG method is 89% over water or CO_2 flooding and 45% over WAG process. The additional recovery resulted from advance of mobility ratio. The comparison of viscosity between WAG and PAG is reported in Figure 5 which describes viscosity of aqueous phase near the injection well. The viscosity obtained from PAG process continuously increases during the injection period of polymeric solution. The betterment of mobility ratio is due to high viscosity of injected polymeric solution and permeability reduction by adsorption. This improvement is indicated by resistance factor in Figure 6. The resistance factor is a ratio of water mobility to polymeric solution mobility. If the viscosity is increased by polymer injection, resistance factor is increased by reduction of polymer mobility [22]. These processes can alleviate viscosity fingering effect in heterogeneous reservoirs. A channeling due to the permeability heterogeneity of this layered system is a dominant factor to reduce sweep efficiency. Figures 7(a) and 7(b) show that improved mobility ratio in PAG process can mitigate viscosity fingering problem and form a stable front.

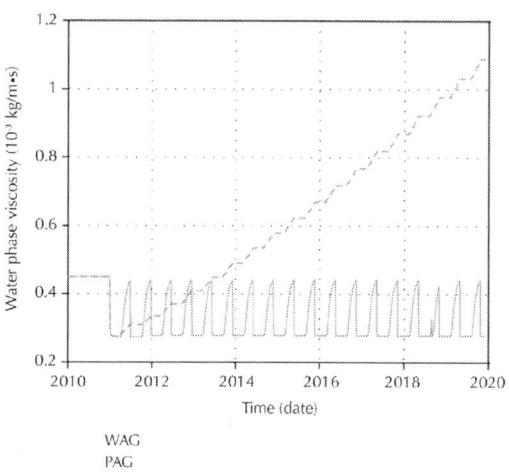

WAG
PAG

Figure 5: Viscosity of aqueous phase in high permeability zone.

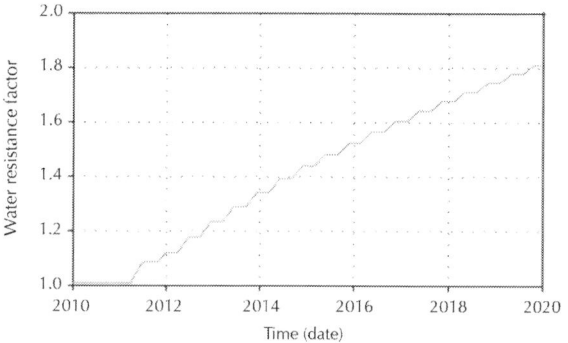

Figure 6: Resistance factor for aqueous phase in high permeability zone.

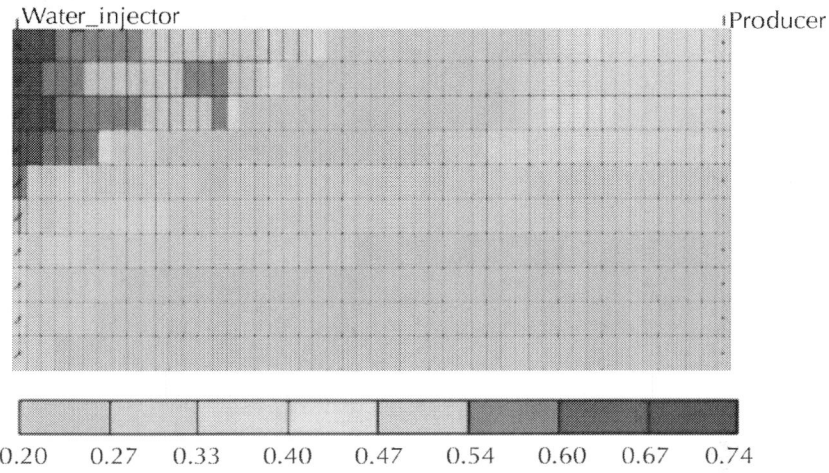

(a)

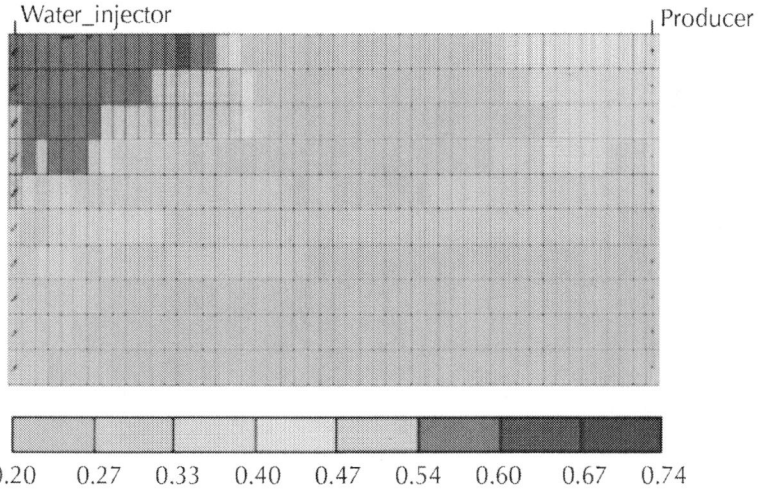

(b)

Figure 7: Water saturation after one cycle of WAG and PAG: (a) WAG process and (b) PAG process.

Figures 8 and 9 represent the water-oil ratio (WOR) and cumulative water production from different processes. According to Figure 8, the results signify that the WOR in the case of PAG process is much lower than those of waterflooding and WAG process during production period, excepting 2014. At this time, WOR of PAG is sharply increased because produced oil rate reaches almost zero due to the temporary blockage with injected CO_2 and polymeric solution (Figure 10). In Figure 9, the PAG process indicates 42% and 12% reduction in cumulative water production compared to the waterflooding and WAG process, respectively. These improvements prove effectiveness for polymer injection which has great potential to reduce aqueous phase mobility.

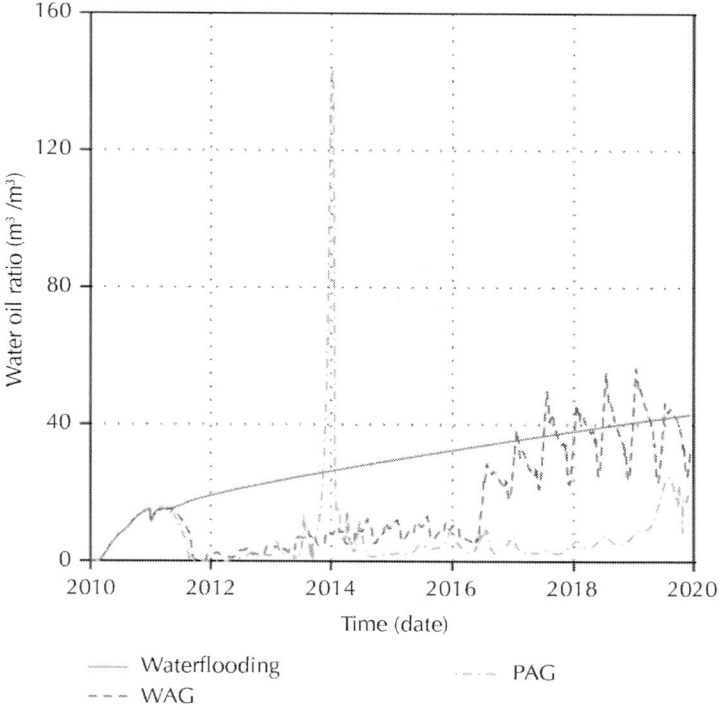

Figure 8: Water-oil ratios for different processes.

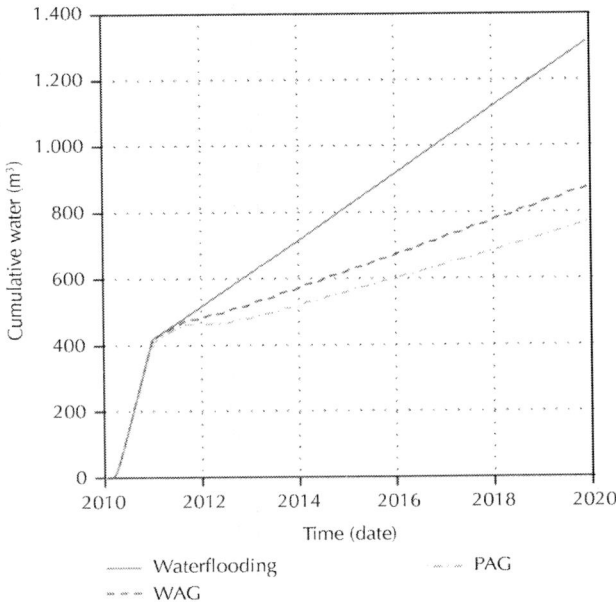

Figure 9: Cumulative water productions for different processes.

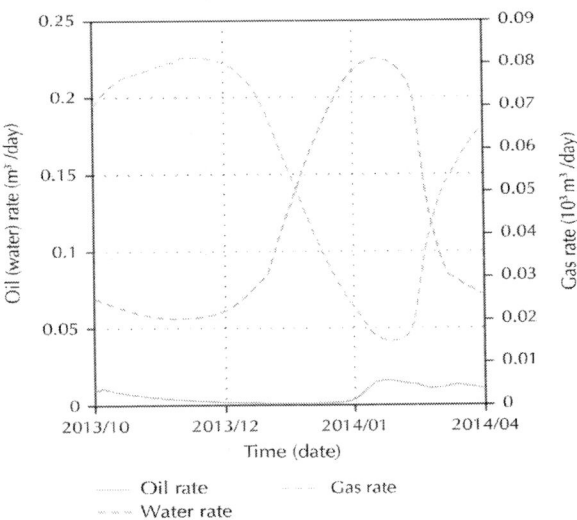

Figure 10: Produced oil, water, and CO_2 rate of PAG process in 2014.

Figures 11 and 12 describe the gas-oil ratio (GOR) and cumulative gas production.In Figure 11, similar problem with WOR existed at 2014. As aforementioned, the same problem at this point results from low oil rate. The amounts of gas productions are $3.4 \times 10^5 m^3$ in CO_2 flooding and $1.6 \times 10^5 m^3$ in WAG and PAG process. PAG process obtains 53% reduction in cumulative gas production compared to CO_2 flooding. GOR of PAG process is significantly lower than that of WAG process. PAG process in heterogeneous heavy oil reservoir attains better performance than WAG does.

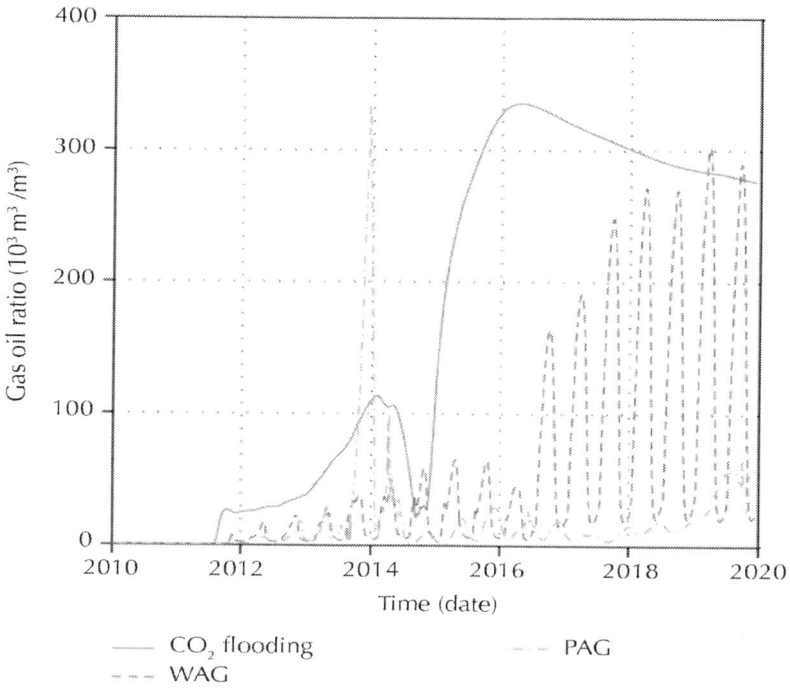

Figure 11: Gas-oil ratios for different processes.

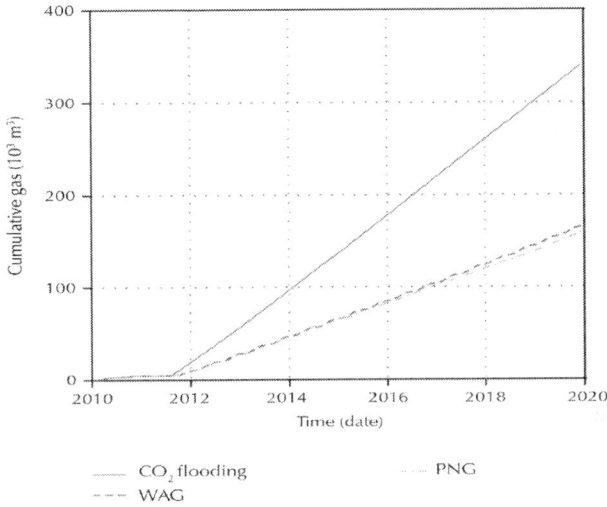

Figure 12: Cumulative gas productions for different processes.

Figure 13 depicts the oil saturation distribution for four processes at the end of production. Average oil saturation is 0.64 in waterflooding, 0.67 in CO_2 flooding, 0.63 in WAG process, and 0.56 in PAG process. In comparison with Figures 13(a), 13(b), and 13(c), Figure 13(d) shows that better recovery efficiency resulted from high sweep efficiency and high displacement efficiency. In PAG process, the reduced permeability contrast due to the preferential adsorption of polymer in relatively high-permeability layers enables water and CO_2 to penetrate into low-permeability layers and the recovery efficiency to be increased.

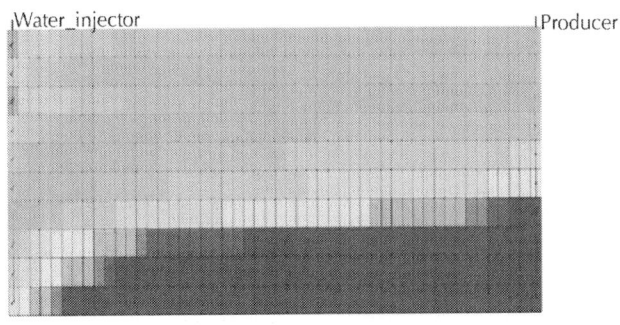

(a)

Water_injector   Producer

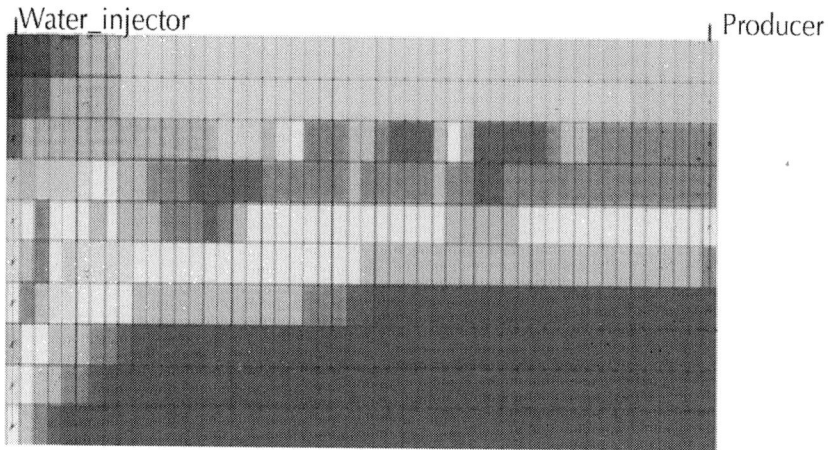

(b)

Water_injector   Producer

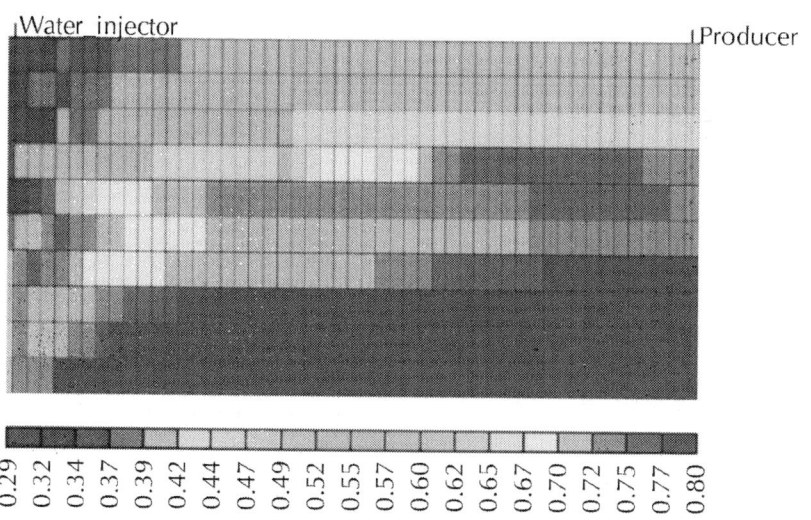

0.29 0.32 0.34 0.37 0.39 0.42 0.44 0.47 0.49 0.52 0.55 0.57 0.60 0.62 0.65 0.67 0.70 0.72 0.75 0.77 0.80

(c)

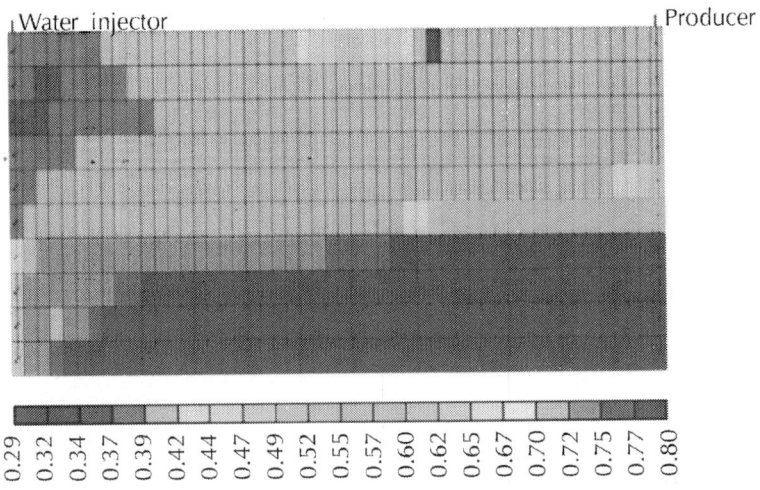

Water injector   Producer

0.29 0.32 0.34 0.37 0.39 0.42 0.44 0.47 0.49 0.52 0.55 0.57 0.60 0.62 0.65 0.67 0.70 0.72 0.75 0.77 0.80

(d)

Figure 13: Oil saturations at the end of simulation: (a) waterflooding, (b) continuous CO_2flooding, (c) WAG process, and (d) PAG process.

PAG Cycle and Ratio

PAG cycle and ratio are general parameters which determine the characteristics of PAG process. The base case is 1:1 PAG ratio and 180 days cycle period. Various PAG cycle periods are applied to compare the oil recovery in the same PAG ratio (1:1). In this PAG process, CO_2 is injected first and polymeric solution follows. The results of these processes are shown in Figure 14. Ultimate recoveries are similar for all cases although increasing points of recovery factors are different. If the respective total amounts of injected CO_2 and polymeric solution are the same in five cases, they have similar efficiencies of sweep and displacement. These results are well matched with those from previous WAG simulation study [23].

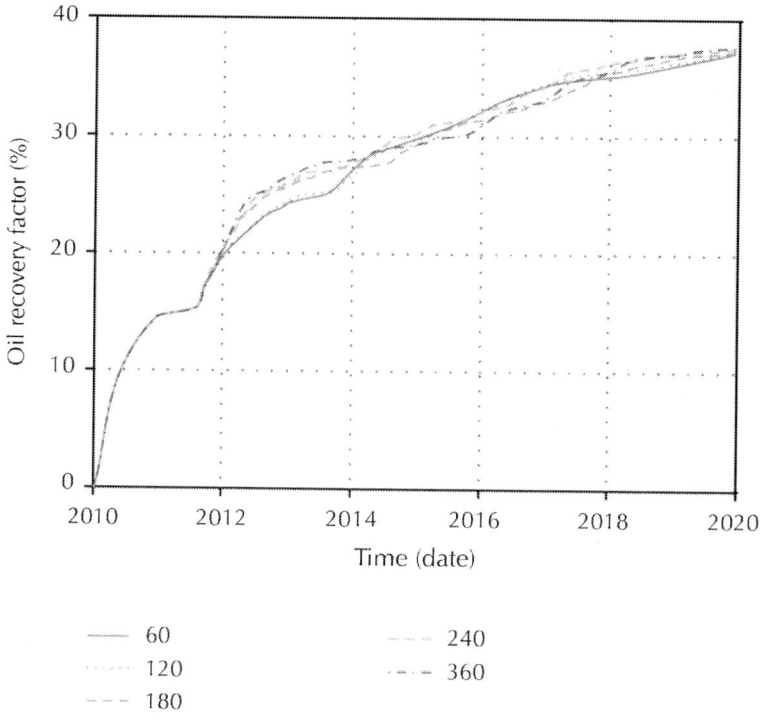

Figure 14: Oil recovery factors at different WAG cycle periods.

Figure 15 is the result of oil recoveries for different PAG ratio processes. Oil recovery factor of 2:1 PAG ratio is 3% larger than that of 1:1 PAG ratio. As a result, injection of more polymeric solution has advantage for oil recovery by increased sweep efficiency. However, too much polymer injection could reduce the oil recovery because mobility of polymer is low and polymer does not reduce residual oil saturation (1:5 PAG ratio case).

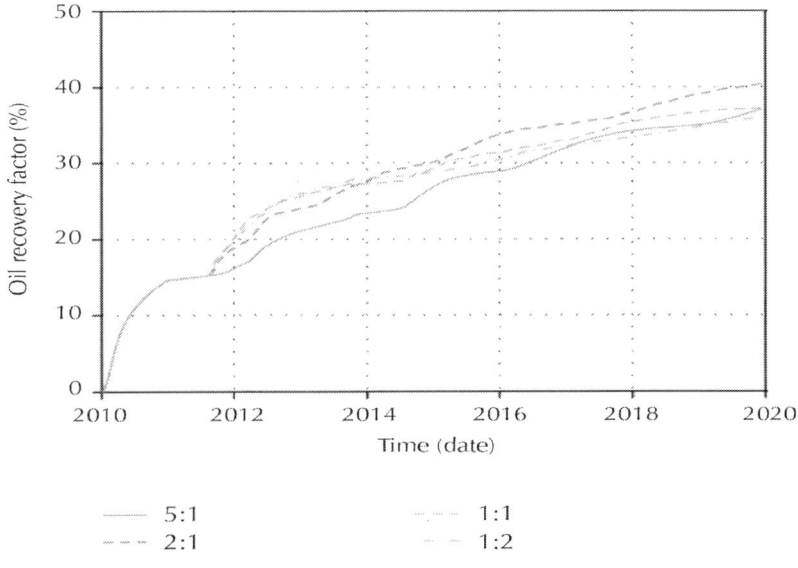

Figure 15: Oil recovery factors at different WAG ratio cases.

CONCLUSIONS

The results of numerical simulation on the flow of polymeric solution with CO_2 in heavy oil reservoir were analyzed. The main challenges to reduce oil recovery are high viscosity of heavy oil and high heterogeneity of reservoir. The polymer-alternating-gas process showed significant advance of recovery efficiency compared with other processes.

- By the control of mobility ratio, the PAG process has better sweep efficiency than those of other processes. The PAG process represented the highest oil recovery factor by 37%. It was 89% higher than results of waterflooding or CO_2 flooding and 45% higher than consequence of WAG process.

- In heterogeneous heavy oil reservoir, water and CO_2 breakthroughs are key factors to reduce the oil recovery in WAG process. In comparison to WAG process, PAG process would decrease the WOR by 12%. Moreover, GOR in PAG process was maintained below GOR of WAG method.

• The cycle time of PAG process did not affect the recovery performance. However, 2:1 PAG ratio could improve the oil recovery factor by about 3% over the base case.

ACKNOWLEDGMENTS

This work was supported by the Energy Efficiency & Resources Core Technology Program of the Korea Institute of Energy Technology Evaluation and Planning (KETEP) granted financial resource from the Ministry of Trade, Industry & Energy, and Republic of Korea (no. 20122010200060).

REFERENCES

1.   T. B. Reid and H. J. Robinson, "Lick creek meakin sand unit immiscible CO_2 waterflood project,"Journal of Petroleum Technology, vol. 33, no. 9, pp. 1723–1729, 1981.

2.   K. Kantar, D. Karaoguz, K. Issever, and L. Varana, "Design concepts of a heavy-oil recovery process by an immiscible CO_2 application," Journal of Petroleum Technology, vol. 37, no. 2, pp. 275–283, 1985.

3.   W. E. Kerr, "Retlaw upper mannvile "V" pool unit experimental carbon dioxide flood," Journal of Canadian Petroleum Technology, vol. 24, no. 1, pp. 275–283, 1985.

4.   A. Spivak, W. H. Garrison, and J. P. Ngugen, "Review of an immiscible CO_2 project, tar zone, fault block V, Wilmington field, California," SPE Reservoir Engineering, vol. 5, no. 2, pp. 155–162, 1990.

5.   W. S. Fong, R. W. Tang, A. S. Emanuel, P. J. Sabat, and D. A. Lambertz, "EOR for California diatomites. CO_2, flue gas and water corefloods, and computer simulations," in Proceedings of the Western Regional Meeting, pp. 159–170, Bakersfield, Calif, USA, April 1992.

6.   K. Issever, N. A. Pamir, and A. Tirek, "Performance of a heavy-oil field under CO_2 injection, Bati Raman, Turkey," SPE Reservoir Engineering, vol. 8, no. 4, pp. 256–260, 1993.

7.   T. D. Ma and G. K. Youngren, "Performance of Immiscible Water-Alternating-Gas (IWAG) Injection at Kuparuk River Unit, North Slope, Alaska," in Proceedings of the SPE Annual Technical Conference and Exhibition, pp. 415–420, New Orleans, La, USA, September 1994.

8.   D. N. Rao, S. C. Ayirala, M. M. Kulkarni, and A. P. Sharma, "Development of gas assisted gravity drainage (GAGD) process for improved light oil recovery," in Proceedings of the SPE/DOE Symposium on Improved Oil Recovery, Tulsa, Okla, USA, April 2004.

9.   B. H. Caudle and A. B. Dyes, "Improving Miscible displacement by gas-water injection," Petroleum Transactions of the AIME, vol. 213, pp. 281–284, 1958.

10.   L. W. Lake, Enhanced Oil Recovery, Society of Petroleum Engineers, Richardson, Tex, USA, 2010.

11.   Y. Zhang, S. Huang, and P. Luo, "Coupling immiscible CO_2 technology and polymer injection to maximize EOR performance for heavy oils," Journal of Canadian Petroleum Technology, vol. 49, no. 5, pp. 27–33, 2010.

12.   S. Majidaie, A. Khanifar, M. Onur, and I. M. Tan, "A simulation study of chemically enhanced water alternating gas (CWAG) injection," in Proceedings of the SPE EOR Conference at Oil and Gas West Asia, pp. 242–250, Muscat, Oman, April 2012.

13.   W. Li, Z. Dong, J. Sun, and D. S. Schechter, "Polymer-alternating-gas simulation: A Case Study," inProceedings of the SPE EOR Conference at Oil and Gas West Asia, Muscat, Oman, March 2014.

14.   S. X. Ning, B. S. Jhaveri, N. Jia, B. Chambers, and J. Gao, "Viscosity reduction EOR with CO_2 & enriched CO_2 to improve recovery of Alaska North Slope viscous oils," in Proceedings of the SPE Western North American Region Meeting, Anchorage, Alaska, USA, 11.

15.   D. Y. Peng and D. B. Robinson, "A new two-constant equation of state," Industrial and Engineering Chemistry Fundamentals, vol. 15, no. 1, pp. 59–64, 1976.

16.   H. Dykstra and R. L. Rarsons, "The prediction of oil recovery by waterflood," in Secondary Recovery of Oil in the United States, pp. 160–175, 1950.

17. S. E. Buckley and M. C. Leverett, "Mechanism of fluid displacement in sands," Transactions of AIME, vol. 146, no. 1, pp. 107–116, 1942.

18. F. F. Craig Jr., T. M. Geffen, and R. A. Morse, "Oil recovery performance of pattern gas or water injection operations from model tests," Transactions of AIME, vol. 204, pp. 7–15, 1955.

19. F. I. Stalkup Jr., Miscible Displacement, Society of Petroleum, Richardson, Tex, USA, 1983.

20. Computer Modelling Group Ltd, "STARS: Advanced Process and Thermal Reservoir Simulator," Calgary, Canada, 2012.

21. J. J. Sheng, Modern Chemical Enhanced Oil Recovery: Theory and Practice, Gulf Professional Publishing, Burlington, Mass, USA, 2010.

22. R. R. Jennings, J. H. Rogers, and T. J. West, "Factors influencing mobility control by polymeric solutions," Journal of Petroleum Technology, vol. 23, no. 3, pp. 391–401, 1971.

23. E. L. Ligero, S. F. Mello, E. O. Muñoz Mazo, and D. J. Schiozer, "An approach to oil production forecasting in a WAG process using natural CO_2," in Proceedings of the SPETT Energy Conference and Exhibition, pp. 109–118, Port of Spain, Trinidad and Tobago, June 2012.

Chapter 4

The Significance of Hydrothermal Alteration Zones for the Mechanical Behavior of a Geothermal Reservoir

Carola Meller and Thomas Kohl

Institute of Applied Geosciences, Division of Geothermal Research, Karlsruhe Institute of Technology (KIT), Adenauerring 20b, Karlsruhe, 76131, Germany

ABSTRACT

Background

The occurrence of hydrothermally altered zones is a commonly observed phenomenon in brittle rock. The dissolution and transformation of primary minerals and the precipitation of secondary minerals affect rocks in terms of mechanics, stress conditions, and induced seismicity.

Methods

The present study investigates commonly observed phenomena of hydrothermal alteration and observations at the geothermal site of Soultz-sous-Forêts, which are related to the occurrence of hydrothermal alteration. Geomechanical observations at Soultz are interpreted on the basis of synthetic clay content logs, which are created from borehole logging data, and which identify clay in hydrothermally altered zones.

Results

It is shown that hydrothermal alteration results in a reduction of the frictional strength of the reservoir rock. Weak zones can act as stress-decoupling horizons, which locally perturb the stress field and affect the evolution of the microseismic cloud. For the first time, it is shown on a reservoir scale that large magnitude seismic events are restricted to unaltered granites, whereas in clay zones, only small magnitudes are observed. It is demonstrated that clay-rich zones foster the occurrence of aseismic movements on fractures.

Conclusions

Secondary mineral precipitation during hydrothermal alteration has a great effect on the geomechanical properties of a geothermal reservoir. The identification of such zones is a first step towards understanding the relation between alteration and mechanical processes inside a reservoir and can help in reducing induced seismicity during hydraulic stimulation of a reservoir.

BACKGROUND

The importance of clay zones for the geomechanical structure and the earthquake mechanics in brittle rock became an important issue in the framework of mitigation studies of natural and man-made disasters (Holmes et al. [2013]). A strong focus was given to hydrothermal alteration in crystalline rock and its effect on mechanical friction. Recent studies on the San Andreas Fault revealed the significant impact of clay inside faults and fractures on their mechanical and hydraulic properties. Faults and fractures are target zones for enhanced geothermal systems (EGS), as they provide pathways for geothermal fluids. In terms of mitigation of induced seismicity, while increasing the permeability of the geothermal reservoir, detailed understanding of hydraulic and mechanical processes of fractured rock is the key for the success of an EGS project.

The Significance of Clay for Geothermal Projects

The development of EGS in low-enthalpy regions like the Upper Rhine Graben in central Europe involves the application of hydraulic stimulation for permeability enhancement in the geothermal reservoir. Mostly located near residential areas, there is a claim for safety and controllability of the geothermal technology from the public. In the past, people were concerned by the occurrence of small perceptible earthquakes, caused by stimulation activities or during operation of geothermal power plants like the magnitude M_L?=?3.4 earthquake in Basel in 2006 (e.g., Häring et al. [2008]), the M_L?=?2.9, M_L?=?2.5, and M_L?=?2.3 in Soultz-sous-Forêts in 2003, 2000, and 2004 (Dorbath et al. [2009]), respectively, or the M_L?=?2.4 and M_L?=?2.7 earthquakes near Landau in 2009 (Groos et al. [2013]). The injection of fluid into the underground changes the effective stress, thus inducing slips on fractures and faults associated with seismic events in brittle rock. In order to predict or even control the seismic behavior of a geothermal reservoir, the geomechanical structures and the associated processes must be known.

In fresh and homogeneous rock, the relation between stress and mechanical failure is commonly described by the Mohr-Coulomb criterion (Scholz [2010]) with flow through fractures to be characterized as sublaminar by the Darcy flow (Sausse [2002]). In geothermal reservoirs, however, the percolation by geothermal brine promotes the formation of hydrothermally altered zones around fluid pathways. The dissolution of primary rock-forming minerals and the precipitation of secondary minerals like quartz, clay, or carbonates change the *in situ* conditions with respect to mechanical strength of the rock. In such zones, simple models might no longer apply, and the reservoir behavior is difficult to assess. Evidence, that simple rock mechanical models no longer account during and especially after the shut-in of hydraulic stimulation, has been only recently highlighted by Schoenball et al. ([2014]) who demonstrated a change in the stress regime during stimulation.

Several studies demonstrate the relation between geomechanics, earthquake characteristics, and the weakness of rocks on a crustal and regional scale. For geothermal projects, however, the geomechanical properties of a reservoir are to be known on a very local scale in the order of several meters. The size of hydrothermal alteration zones can range from millimeters to several kilometers. In order to characterize a geothermal reservoir and to assess its geomechanics, it is important to understand the significance of such alteration zones. Therefore, it is necessary to know and to understand, if and how large-scale geomechanical rules and observations can be transferred to the reservoir scale.

The present paper conducts an investigation on the significance of hydrothermal alteration in the granite of the geothermal site in Soultz-sous-Forêts (France) and the change of its mechanical parameters. The basis of the analyses is synthetic clay logs, which are created from spectral gamma ray logs using a technique introduced by Meller et al. ([2014a]). These logs are indicative of the occurrence of clay-bearing fractures along the boreholes. The newly derived results are investigated under the light of the existing geomechanical interpretation, which is summarized in the 'Current state of research on the role of clay in fault zones' subsection.

Current State of Research on the Role of Clay in Fault Zones

Evidence for the role of clay as zones of weakness or some kind of lubricant on faults promoting aseismic movements has been described by Schleicher et al. ([2006]), Dolan et al. ([1995]), and Wu et al. ([1975]). Clay minerals are a characteristic of creeping faults with rates of up to 30 mm/a assumed for the San Andreas Fault (Chang et al. [2013]). Studies on the slipping behavior of the San Andreas Fault, however, suggest that it is not merely creeping but it rather consists of creeping patches, which build up stress on patches with high friction. If the stress is large enough, these high friction patches rupture and cause seismic events (e.g., Chang et al. [2013]). This theory is supported by the work of Amelung and King ([1997]) who observed a continuous earthquake activity on creeping faults. A major result of their study is that creep and earthquakes are not two separated phenomena, but two processes which go hand in hand. This has been reported earlier for numerous faults and continental margins (e.g., Brune [1968]; Voisin et al. [2004]; Mulargia et al. [2004]). Recent studies at the geothermal site in Soultz-sous-Forêts revealed similar mechanisms during shear movements on faults. Bourouis and Bernard ([2007]) observed in the data of seismicity induced during GPK1 stimulation repeated shear movements on fault asperities surrounded by creeping zones. Schmittbuhl et al. [2014] observed in the laboratory experiments a close relationship between seismic and aseismic movements on faults and conclude that aseismic processes can drive seismicity, almost independent from fluid pressure. The triggering of seismic events by creep movements is an important issue for EGS and needs to be considered for the mitigation of large seismic events (Figure 1).

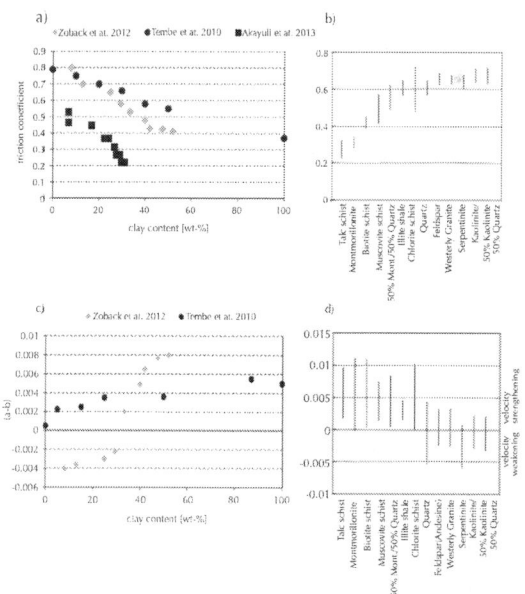

Figure 1: Friction coefficient and parameter (*a-b*) for different rock types. (a) Friction coefficient for rocks and a soil with different clay contents. Increas-

ing clay content reduces the frictional strength. (b) Range of measured friction coefficients for different rock types (data from Ikari et al. [2011]). (c) (a-b) parameter for rocks and soil with different clay contents and (d) (a-b) for different rock types. With increasing clay content, the friction coefficient is reduced and (a-b) increases. High clay content and low friction result in velocity-strengthening behavior. Rock types with low friction show velocity-strengthening behavior. Experimental data from Zoback et al. ([2012]) is derived from measurements on shale, and Tembe et al. ([2010]) measured artificial quartz-illite samples. Akayuli et al. ([2013]) did experimental studies on soil.

Clay minerals, which are a main product of hydrothermal alteration (e.g., Meunier [2005]; Velde[1995]), sometimes have very low friction coefficients of approximately 0.3 (e.g., Morrow et al.[1992] and references herein, and c.f. Figure 1b). The frictional properties of clay minerals, however, strongly depend on their structure and water content. Therefore, it is not easy to estimate the frictional properties of clay-filled faults (Moore and Lockner [2007]). Many studies have been conducted on the relationship between the nature of fracture fillings and fault mechanics. Zoback et al. ([2012]) and Kohli and Zoback ([2013]) investigated the relationship between clay content and the mechanical friction of shale gas reservoir samples under wet conditions. They observed a linear decrease of the friction coefficient with increasing clay content (Figure 1a) from 0.8 with 10 wt.% clay to 0.4 at approximately 50 wt.% clay. Similar results have been obtained by Tembe et al. ([2010]) for artificial clay gouge samples of quartz and illite and for natural soil samples tested by Akayuli et al. ([2013]). The friction coefficients they measured for different clays vary and are much lower than those of other minerals like quartz or feldspars (Figure 1b).

The rupture behavior of a fault from the Dieterich-Ruina constitutive model (Ruina [1983]; Dieterich[1978]) describes the frictional evolution of a fault for different sliding velocities with the material parameter (a-b) representing the difference in steady-state friction. It indicates stable sliding of fault surfaces during slip if (a-b)?>?0 or unstable sliding if (a-b)?<?0. The synonyms for stable and unstable sliding are velocity-strengthening and velocity-weakening behavior, respectively. The effect of clay on the rupture behavior of faults has been studied by many laboratory experiments. Ikaris et al. ([2011]) found experimental evidence for the relationship between the weakness of rocks and their frictional stability: rock samples with a low friction coefficient

show velocity-strengthening behavior, whereas samples with high friction coefficients show velocity-weakening behavior (Figure 1). This indicates the occurrence of brittle failure only on rocks with high friction coefficients. Zoback et al. ([2012]) observed experimentally on shale gas samples that faults with clay contents higher than 30% slide stable (i.e., (a-b)?>?0), whereas faults with a lower clay content slip unstable (i.e., (a-b)?<?0, Figure 1c). They reasoned that such clay-rich faults slide aseismically, whereas the faults with lower clay contents produce microseismic events. The dataset of Tembe illustrates a dependence of (a-b) of illite-quartz samples on the illite content. For these samples, no velocity-weakening behavior is observed. The reason for this is that quartz can behave both velocity strengthening and velocity weakening, and under the experimental conditions, it was velocity-strengthening (a-b)?>?0 (Figure 1c), but nevertheless the effect of the clay proportion of the samples on (a-b) is significant.

As the frictional properties of rocks determine their slipping behavior, a correlation between the weakness of the rocks and the occurrence of large and small earthquakes is expected. The so-called b-value, which is derived from the Gutenberg-Richter law (Gutenberg and Richter [1954]), describes the proportion of small relative earthquakes to large ones. A b-value of 1 represents a logarithmic relationship between the magnitude of events and their frequency, whereas b-values?>1 reflect an increased number of small earthquakes. High b-values are expected in areas where no large differential stress can build up. Schorlemmer et al. ([2005]) compared the results of numerous earthquakes from different settings and of laboratory data. They found that the b-value differs systematically with the faulting regimes. The highest b-values are found in normal faulting regimes (up to 1.2), whereas the lowest b-values occur in thrust events (as small as 0.6), and strike-slip events are in between. Based on the stress prevailing in the respective regimes, Schorlemmer et al. ([2005]) concluded that the b-value inversely correlates with differential stress levels. This was also confirmed by laboratory experiments performed by Amitrano ([2003]) who observed a decreasing b-value with increasing differential stress. Creeping fault sections show very high b-values of around 1.3 (Schorlemmer and Wiemer [2005]). Based on these results, the occurrence of small events and aseismic movements in strongly altered and fractured areas is expected rather than large earthquakes. This assumption has also been proposed by Heinicke et al. ([2009])

who investigated the correlation between hydrothermal alteration and the occurrence of earthquake swarms. They observed in the Vogtland region of northwestern Bohemia that in addition to increased pore pressure and shear stress, the mechanical weakening of the rocks and the dissolution of fracture walls play an important role for the evolution of earthquake swarms. Interestingly, the maximum magnitude of such earthquake swarms is limited to 5 (Heinicke et al. [2009]), which supports the theory of only small earthquakes occurring in regions with rocks of low friction coefficients. When analyzing b-values, one has to consider that this value is affected by numerous parameters, not least by the way it is computed. Besides the strength of the rock, the main affecting parameters are the stress field, the focal mechanism of the earthquakes, and the presence of large geologic structures (Scholz [2010] and references herein). In geothermal reservoirs, large variations of b-values in time and space have for example been observed by Bachmann et al. ([2012]). They calculated the b-value for the time period during injection and after injection. The b-values varied from 1.58 during injection to 1.15 after injection, which represents a larger proportion of small earthquakes during injection.

Dorbath et al. ([2009]) calculated a b-value of >1.2 for the stimulations of the well GPK2 at Soultz, whereas for the well GPK3, which is a maximum 500 m away from GPK2, was determined to be 0.9. They related this behavior to the presence of large fault zones in the vicinity of the well, which dominate their seismic evolution.

The Soultz Geothermal Site

The European geothermal project of Soultz-sous-Forêts (France) targets a geothermal anomaly at the western border of the Upper Rhine Graben. Five wells have been drilled to a maximum depth of 5 km. Three of these wells are currently used for operation with two wells as injectors (GPK3 and GPK4) and one producing well (GPK2) (Genter, personal communication). The upper geothermal reservoir is hosted by a porphyritic Hercynian monzogranite (Figure 2a), overlain by 1.4 km of Mesozoic sediments. The lower reservoir lies in fine-grained two-mica granite, which can be encountered at depths greater than 4.6 km. The pluton has been affected by the Upper Rhine Graben tectonics, which caused the formation of large sets of faults and fracture zones. These faults and fractures are the main pathways for circulating fluids and are thus responsible for the permeability of the rock (Genter and Traineau [1992]). Paleo-circulation of meteoric fluids from the Graben shoulders led to pronounced alteration of the Soultz granite. The first pervasive alteration affected the whole granitic matrix (Figure 2b) but had no effect on the structural properties of the granite. It involved the formation of mainly chlorite and hematite. A subsequent vein alteration event significantly changed the granite structure (Figure 2c). During this alteration event, primary minerals were dissolved, and secondary minerals precipitated (Schleicher [2005]). Alteration halos developed enfolding the zones around fractures affected by hydrothermal alteration. These halos can be several tens of meters thick and are characterized by the transformation of mainly silicates and the precipitation of secondary clay minerals, quartz, carbonates, sulfates, and iron oxides (Genter and Traineau [1996]). The dominating clay minerals of the vein alteration are several generations of illites and smectites, and minor tosudite and chlorite (e.g., Bartier et al. [2008]).

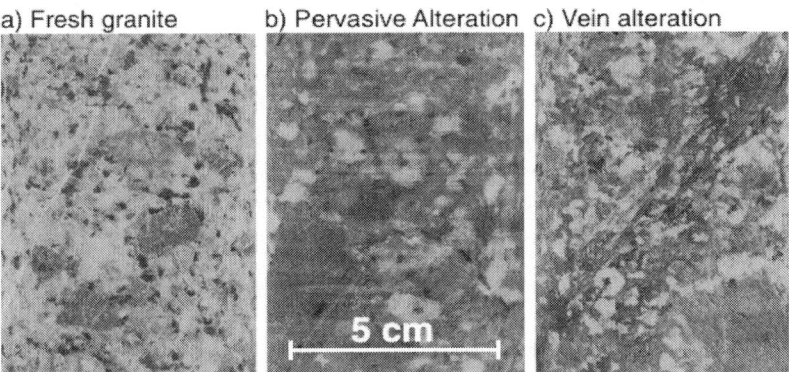

Figure 2: Core pictures of different facies of the Soultz granite. (a)Fresh granite without evidence of hydrothermal alteration. (b) Pervasive alteration with mainly formation of hematite and no structural influence. (c)Vein alteration with dissolution of silicates and precipitation of clay minerals.

The sealing of fractures by secondary minerals and the transformation of silicates into clay minerals affected the hydraulic and mechanical properties of the rock (Valley and Evans [2003]; Charléty et al.[2007]), whereas the details of such processes are still subject to extensive research. Bartier et al. ([2008]) highlighted for example the importance of clay mineralogy for the permeability of the Soultz granite, which is reduced by illite precipitation but enhanced by tosudite precipitation. Ledésert et al. ([2010]) highlighted the complexity of processed linked to porosity/permeability formation and decrease by the dissolution and transformation of primary minerals and the formation of new minerals. The type and structure of clay minerals are not only important for the evolution of porosity and permeability but also for the shearing properties of a fault filled with clays.

The variation of hydro-mechanical properties of the rock with different alteration types and grades makes it important to first detect alteration zones and, second, to understand their significance for the performance of a reservoir (Figure 2).

METHODS

The basis for the rock mechanical studies are neural network-derived synthetic clay content logs (SCCL), which present the clay content along the borehole in a semi-quantitative way with five groups of increasing clay content. In sedimentary rocks, clay minerals can be easily identified from peaks in spectral gamma ray (SGR) logs. In crystalline rock in contrast, apart from clays, numerous other minerals contain radioactive isotopes, which makes it difficult to identify clay minerals on SGR logs. Therefore, a neural network is used, which makes it possible to identify different signal patterns on logging data and to localize the clay-bearing zones. The resolution of the resulting SCCLs is on the scale of decimeters.

The neural network for the creation of the synthetic clay content logs uses a self-organizing map working with a Kohonen algorithm (Kohonen [1984]). The principle of this procedure is the grouping and indexing of patterns according to their spatial distance from each other. Each combination of $n$logs represents a vector in n dimensions. The location of these vectors in n dimensions determines their assignment to the nodes of a two-dimensional self-organizing map. Thus, their dimension is reduced, which makes it easier to classify them. The number of classes can be defined according to the desired resolution. The network is trained using supervised learning, which means the grouping and classification of the nodes based on reference data. These reference data are used to teach the neural network how different parameters are correlated. For this study, the reference data consists of spectral gamma ray logs and a fracture density log. A log representing the density of clay-filled fractures was derived from core investigations of the EPS1 reference well and served as a template for the classification of the nodes. For the deep wells GPK1 to GPK4, the fracture density log could be computed on the basis of fractures identified on borehole image logs. The resulting SCCLs (Figure 3) semi-quantitatively represent the density of clay-filled fractures along the boreholes with five groups. The major flow paths through the granite are marked by dashed lines. Group 1 represents the group with the lowest clay content, and group 5 represents the highest clay content. The comparison with reference data has shown that only approximately 10% of the logs deviates more than 1 SCCL group from real data and the vertical resolution of the logs is between 10 and 50 cm depending on the resolution of the SGR logs, from which they are derived.

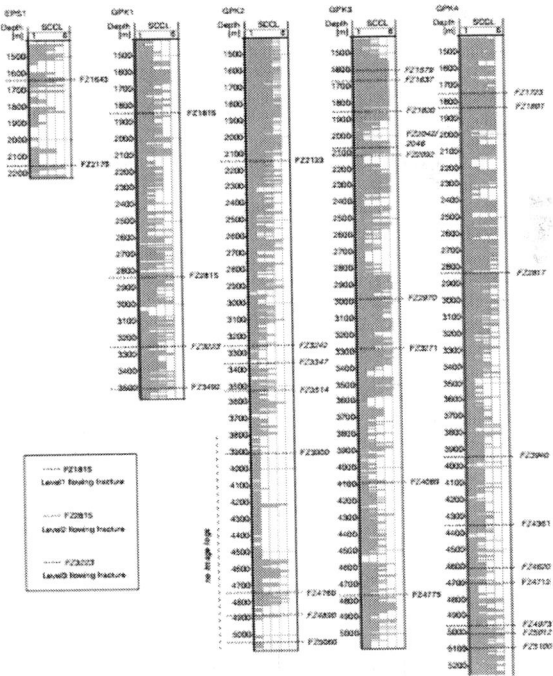

Figure 3: SCCL for the wells GPK1 to GPK4. The SCCL is created on spectral gamma ray and the corrected fracture density log. FZXXXX denotes fracture

zones after their depth of occurrence. Level 1 flowing fractures: major fracture zones with significant mud losses during stimulation, primary permeable. Level 2 flowing fractures: >20% water losses during stimulation. Level 3 flowing fracture: small fracture zones with <20% water loss (classification after Dezayes et al. [2010]). Higher SCCL indicates higher clay contents with SCCL?=?1 representing no clay and SCCL?=?5 representing the maximum density of clay-filled fractures.

The SCCLs allow discriminating between zones of high and low clay contents. Whereas the upper parts of all wells are characterized by high SCCL values, representing the paleo-alteration surface, the lower parts are very different for the five wells. Intervals with high SCCL are mostly found around fractures, which have been identified as permeable on flow logs, but hydrothermal alteration also occurs away from such fractures. However, not all permeable fractures are located in altered zones. This might be due to the fact that extreme alteration leads to a clogging of fractures with clay minerals, thus reducing its permeability (Sausse [2002]). The actually flowing fractures might not have been permeable in the past, which prevented the surrounding rock from being hydrothermally altered. Increased clay content is seen at the bottom of the wells below 4,600 m at the transition between the porphyritic and the two-mica granite. For details of this neural network method, the SCCLs, and the calibration of the logs by magnetic mineralogical investigations refer to Meller et al. ([2014a]; [2014b]) (Figure 3).

For the deep wells in Soultz, no core material is available. Therefore, petrophysical and geologic parameters can only be derived from borehole measurements and seismicity catalogs. This study is mainly based on breakout and fracture analyses conducted on borehole image logs and on a catalog of seismic events recorded during hydraulic stimulation. Borehole breakouts are enlargements and elongations of a borehole in a preferential direction and are formed by spalling of fragments of the wellbore during drilling. They generally form parallel to minimum horizontal stress, and their formation is facilitated in weak wall rocks (Babcock [1978]). Their analysis can therefore provide information about the orientation of the stress field and on the mechanical properties of the penetrated rock. Seismic events induced during stimulation are an indication of structures in the geothermal reservoir. Their analysis provides indications about the stress state, fracture orientation, rock mechanics, and fluid pathways.

RESULTS AND DISCUSSION

Impacts of Hydrothermal Alteration on Rock Mechanics

Due to their preferential formation in weak rocks, a cumulative occurrence of breakouts in altered zones in the Soultz granite could be indicative of the weakness of alteration zones. This theory was investigated on the basis of breakouts, identified by Sahara et al. ([2014]). They analyzed breakouts in the deeper part of the well GPK4. On Figure 4, a total of 2,440 breakouts from Sahara et al. ([2014]) together with the SCCL for this depth interval are illustrated. A correlation between clay content and the appearance and size of the breakouts is obvious. Whereas the density of breakouts is high in clay-rich intervals, depth sections without clay are characterized by an absence of breakouts, as for example at 4,180, 4,480 to 4,590, and 4,730 to 4,800 m (BA in Figure 4). Obviously, the occurrence of breakouts is strongly related to the presence of hydrothermally altered zones. This suggests that hydrothermal alteration weakens the rock and thus promotes the formation of breakouts. Upon the transition from the porphyritic granite to the two-mica granite at around 4,800 m, the breakout density clearly increases (2M in Figure 4). The cumulated occurrence of breakouts at this transition might originate from the mechanical contrast between the two granites. Such mechanical contrast occurs also at the transition between fresh granite and strongly altered granite.

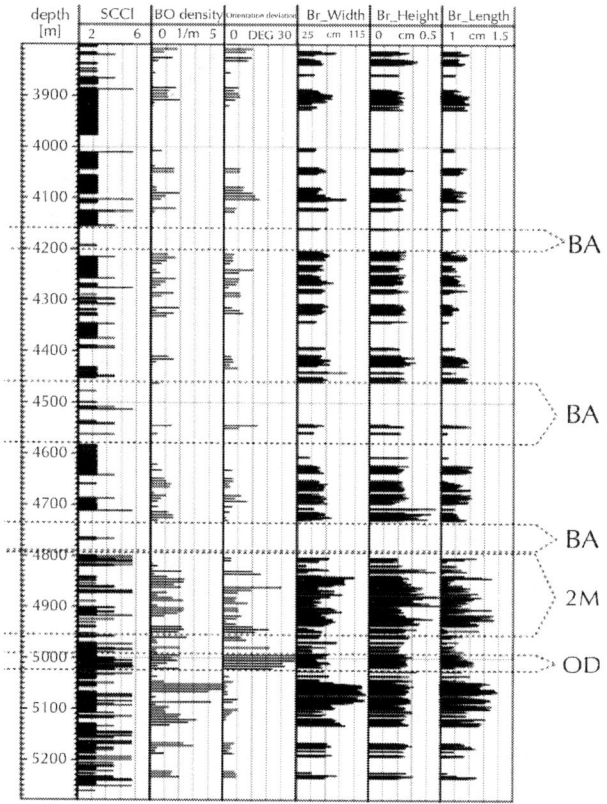

Figure 4: SCCL 2 to 5 of the well GPK4 compared to the breakout density, orientation deviation, and breakout dimensions. There is a clear correspondence between breakout appearance and size and the SCCL (breakout data from Sahara, personal communication). Intervals without clay are characterized by an absence of breakouts (BA). On the transition between standard granite and two-mica granite at around 4,800 m, the orientation of the breakout changes (2M). This might be attributed to the mechanical contrast between the two granitic rock masses. A clay-rich interval around 5,000 m depth coincides with a large orientation deviation of the breakouts (OD).

The correlation of the clay zones with the occurrence of breakouts demonstrates their geomechanical significance. However, it is in contrast to present studies, which seem to identify rather high friction coefficients for the Soultz granite. By applying the Mohr-Coulomb failure criterion using effective stresses on the fractures, Cornet et al. ([2007]) obtained a minimum friction coefficient of 0.81. Evans et al.

([2005a]) found that fractures in highly altered zones are surprisingly strong despite the presence of illite and ascribed this behavior to their internal architecture of intact rock bridges and jogs between weak zones. According to Byerlee ([1978]), rock samples with precut fault surfaces have a uniform friction coefficient of 0.85 and no cohesion at normal stress below 300 MPa, independent from the rock type. However, Byerlee ([1978]) observed that fractures filled with clay minerals are an exception from his law and have much lower friction coefficients. A gradual decrease in frictional strength with the addition of clay has also been observed in triaxial measurements of the Berea sandstone under high temperature and high pressure conditions. Here, a sharp drop of the friction coefficient occurred at clay contents of approximately 50% (Takahashi et al. [2007]). In these experiments, clay minerals inside fractures were observed not only to weaken faults but also to stabilize their sliding behavior. Crawford et al. ([2008]) performed experiments of quartz-kaolinite mixtures of different proportions and compared their strength. They also observed a reduction of frictional strength with an increasing clay fraction. Ikari et al. ([2009]) obtained low friction coefficients of fault gouges rich in phyllosilicates (Figure 5).

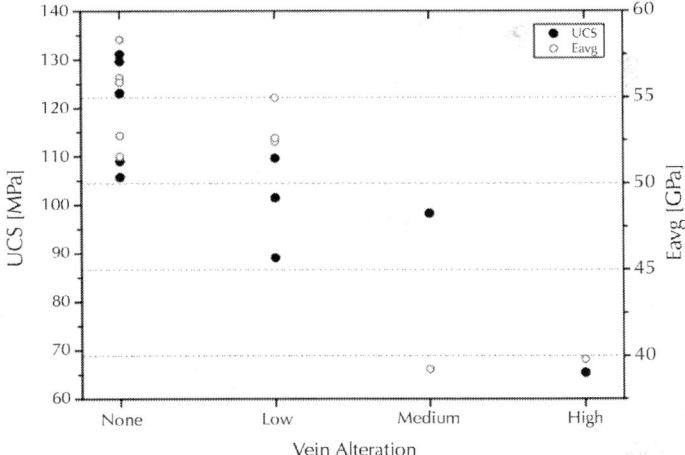

Figure 5: E-moduli and uniaxial compressive strength (UCS) of samples with different vein alteration grades. Data is taken from Valley and Evans ([2003]) who measured UCS and E-modulus for samples with different alteration grades. Alteration intensity is inversely correlated with E-modulus and UCS.

These experimental results are in agreement with the breakout observations at Soultz, which indicate weakness of the hydrothermally altered zones, but which are in contrast to the high minimum friction coefficient of 0.81 determined by Cornet et al. ([2007]) for the whole granitic rock mass. It is therefore assumed that hydrothermal alteration causes a variation in the frictional properties of the Soultz granite on a meter scale with higher frictional strength in unaltered rock and a lower frictional strength in altered rock.

Elastic properties of the Soultz granite have been experimentally studied by Valley and Evans ([2003]). They selected samples of different alteration grades from the EPS1 core and measured the uniaxial compressive strength (UCS) of the core pieces. Furthermore, they measured the S- and P-wave velocities of the samples in order to determine their E-moduli. They found an inverse correlation between alteration grade and UCS and the E-modulus of the samples (Figure 5). From the results of this study, it is expected that the highly altered clay zones affect the frictional properties in Soultz and the friction coefficient is not uniform but is lowered by hydrothermal alteration.

Recent researches showed that a characteristic of such weak zones is that they can fail at low stress levels, as it is for example observed on a large scale on the San Andreas Fault in a strike slip regime, whose slip direction deviates 70° from the maximum horizontal stress (e.g., Boness and Zoback[2006]), the Zuccale normal fault on Elba (e.g., Smith et al. [2007]) or some normal faults at the eastern side of the Sea of Japan (e.g., Sibson [2009]; Faulkner et al. [2010]). If such observations can be transferred to the reservoir scale, hydrothermally altered zones might fail at lower stress levels than the surrounding intact rock mass. This is especially important in terms of hydraulic stimulation, as weak faults could shear at much lower stimulation pressure than unaltered rock and influence the evolution of induced seismicity (Figure 4).

Impacts of Hydrothermal Alteration on the Stress Field

Clay layers inside rock masses give rise to large contrasts of mechanical properties. In contrast to intact crystalline or sedimentary rock masses, weak clay-rich zones cannot establish large differential stress (Zoback and Harjes [1997]).

The stress field in Soultz has been thoroughly investigated by many scientists (Valley and Evans[2010]; Cornet et al. [2007]; Rummel [1995], and references herein). This resulted in detailed knowledge of the magnitude and orientation of the principal stress components with depth. Based on the analysis of wellbore failure and hydraulic data as well as microseismic data, a linear stress model has been established for the Soultz reservoir (Valley and Evans [2007]; Cornet et al. [2007])

$$S_v = -1.3 + 0.255z$$
$$S_H = 0.98(-1.3 + 0.255z)$$
$$S_h = -1.78 + 0.01409z$$
$$P_p = 0.9 + 0.0098z$$

with P_p the pore pressure and z the depth in meters.

The S_H orientation is approximately north-south, and the vertical stress S_V is equivalent to the overburden. However, in inhomogeneous rock masses with changing mechanical properties, the magnitude and orientation of the stress field change at the transition between layers of different mechanical strength. The Soultz granite is very heterogeneous due to its porphyritic structure, its lithological variations, hydrothermally altered zones, and the profound fracturing. Borehole breakouts generally form in the direction of the minimum horizontal stress and are therefore useful indicators of the orientation of S_h and S_H. An analysis of borehole breakouts can give evidence about local stress variations. The high resolution of the SCCLs in the order of decimeters for the first time allows a detailed analysis of the indications for stress field variations at Soultz on the basis of breakouts. In the following section, the occurrence of breakouts and their orientation is interpreted on the basis of the SCCLs.

Evidence for a change in the direction of the principal stresses can be found in breakout data from Sahara et al. ([2014]) in the well GPK4. The transition from porphyritic to two-mica granite at around 4,800 m (2M in Figure 4) is not only characterized by an increased breakout density but also by a deviation of the mean breakout orientation. As breakouts generally form in the direction of the minimum principal stress, an orientation deviation of the breakout could be an indicator for a rotation of S_h. Another excursion of the breakout orientation is observed at a depth of 5,000 m (OD in Figure 4). This deviation coincides with a very clay-rich interval. This clay-rich interval might act as a small stress-decoupling horizon, which rotates the principal stress. According to the data of Sahara et al. ([2014]), stress rotations in clay-rich intervals can be as large as ±30° from the mean orientation. Deviations of breakout orientations in hydrothermally altered zones have also been observed by Valley ([2007]) in the wells GPK3 and GPK4. He observed major stress perturbations occurring at depths of 2,000 and 4,700 m. The SCCLs show that these stress perturbations coincide with the occurrence of clay-rich intervals related to large-flowing fracture zones (FZ2123, FZ4760, FZ4775, Figure 3), but it cannot be ruled out that they could also be caused by the presence of the fracture zones as it was, for example, observed by Valley ([2007]). The stress variation at approximately 4,700 m is probably caused by the contrast in the elastic moduli of the rock between the standard porphyritic granite and the two-mica granite at this depth and the increased clay content (Figure 6).

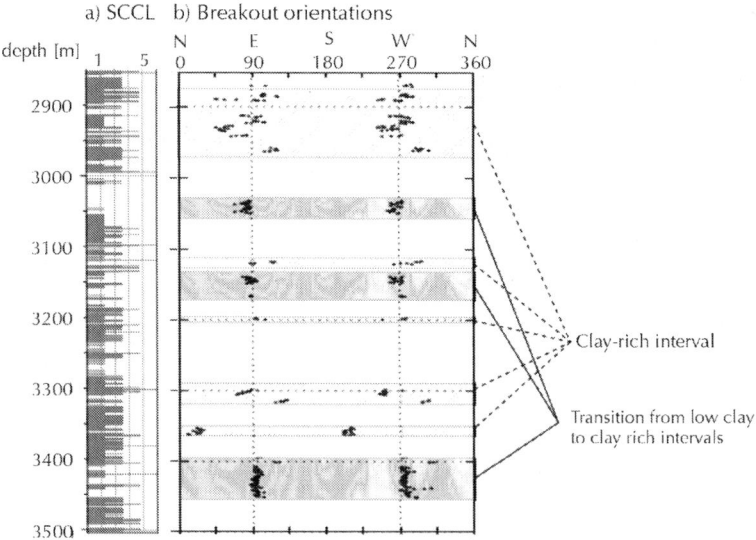

Figure 6: SCCL and breakouts for the well GPK1. A mean breakout orientation of 90°? ± ?19° was determined from 498 measurements (data from Valley and Evans [2000]). The occurrence of clay-rich intervals (shaded panels) coincides with an accumulation of borehole break-

outs. At the transition between fresh and altered granite (patterned panels), the breakout orientation rotates.

Valley and Evans ([2000]) analyzing breakouts in the well GPK1 between 2,840 and 3,510 m found an increased breakout concentration at the top of this interval. This agrees with the occurrence of a clay-rich interval in this section indicated by high SCCL (Figure 6). The mean S_H orientation determined from breakouts is 0°?±?19°, which is in agreement with the mean orientation of the structures of the microseismic cloud. Excursions of the mean breakout orientation occur in the intervals 2,890 to 2,950 and 3,300 to 3,350 m, which are characterized by high SCCL values. The occurrence of breakouts in GPK1 between 2,960 and 3,500 m is not only restricted to high-clay zones but high breakout-densities as, for example, at 3,000 to 3,050 m or at 3,400 to 3,450 m depth also occur, when a depth interval without or with very little clay is followed by a very clay-rich interval. Here, the contrast of elastic moduli of the two depth intervals might cause a cumulating appearance of breakouts. This might also be represented in the different orientations of the microseismic cloud in the depth intervals 2,700 to 2,900 m, where it is oriented north-south, and 3,200 to 3,600 m, where its azimuth is 145° to 160° (Cornet et al. [1997]). Cornet and his colleagues ([1997]) linked this orientation deviation to the higher pore pressure above 2,900 m, but it could also be related to the presence of clay-rich zones. Such clay-rich zones could also lead to increased pore pressures (Wu[1978]).

Similar analyses have been conducted by Langenbruch and Shapiro ([2014]) who investigated stress states in boreholes from different regimes. Based on sonic logs, they created a model of the *in situ*elastic moduli to calculate the spatial distribution of *in situ* stress within a rock mass. Their large spatial variations of the stress regime suggest that linear stress models are not sufficient for Coulomb failure within a rock mass. Economides et al. ([1989]) observed that within sedimentary formations, the vertical gradient of the minimum horizontal principal stress does not vary linearly with depth. The authors found that elastic heterogeneity has a significant influence on stress magnitudes, which vary by up to more than ±20% of the externally applied stresses. Cornet and Roeckel ([2012]) observed this phenomenon in limestone layers of the Paris Basin and in the North German Basin. They saw that the local stress magnitudes are not linearly increasing with depth, and they saw variations of approximately 15° in the stress directions. In contrast to Langenbruch and Shapiro ([2014]) and Economides et al. ([1989]), they assume that the stress magnitudes are controlled by the creeping characteristics of the various layers rather than by their elastic characteristics (Cornet and Roeckel[2012]).

The change of the local stress field in magnitude and orientation has previously been described for large fracture zones (e.g., Brudy et al. [1997]). In the San Andreas Fault, for example, a stress rotation of approximately 28° with respect to the stress field of the rigid crust has been measured (Chéry et al. [2004]). Cornet and Roeckel ([2012]) identified soft layers as decoupling layers introducing decoupling of stress fields in the layers above and below these layers. This was also observed by Meixner et al. ([2014]) who documented a rotation of the maximum horizontal stress in different facies along the Bruchsal geothermal wells (c.f. Figure seven in his article).

However, in those studies, stress field variations are only observed on large scales of several kilometers. The analysis of breakouts on the basis of SCCLs provides indications that changes of the stress field both in magnitude and orientation of the principal stress can also be induced by small-scale soft alteration zones on the meter scale as observed in geothermal wells. Taking these observations into account, it is obvious that the estimation of mechanical properties on the basis of a linear stress field can only provide far field values, especially for zones, where the SCCL is high. So, in addition to the frictional parameters, the exact orientation of the stress field has to be constrained in hydrothermally altered zones in order to be able to assess their mechanical characteristics.

The Impacts of Hydrothermal Alteration on Induced Seismicity

At Soultz, 20 hydraulic and chemical stimulations have been performed and large catalogs of seismic events are available (Genter et al. [2010]). During hydraulic stimulation, large amounts of water are injected into the geothermal reservoir in order to increase the pore pressure prevailing in the reservoir rock. If the pressure increase is large enough to overcome the frictional stability of fractures, shear movements are induced, which can be observed by the occurrence of microseismic events. A detailed summary of the background of hydraulic stimulation can, for example, be found in Economides et al. ([1989]) or Majer et al. ([2007]).

The parameters influencing the evolution of induced seismicity like the pressure of the fluid, the ambient stress field, the orientation of fractures, hydraulic properties, and the frictional characteristic of rock can be affected by hydrothermal alteration. Herein, the relation between hydrothermal alteration and induced seismicity at Soultz is investigated.

Except for some new fractures, which are created during hydraulic stimulation at Soultz by hydrofracturing (Cornet [2012]), seismicity in the Soultz reservoir is mainly restricted to shear movements on existing geological structures, which can be observed during all stimulations performed on the Soultz wells (e.g., Evans et al. [2005b]; Fabriol et al. [1994]; Dorbath et al.[2009]). However, it is not clear why some structures are seismically more active than others. According to the Mohr-Coulomb failure criterion, the most important factor affecting the shearing behavior of a fracture or fault is its orientation relative to the ambient stress field. Favorably oriented fractures lie (sub)-parallel to the maximum principal stresses and can thus be easily sheared. The focal mechanisms at Soultz (cf. Figure one in Schoenball et al. [2012]) indicate that some fractures produce seismic events upon hydraulic stimulation unless they should be stable according to the Mohr-Coulomb failure criterion. A possible reason for that could be a very low shear strength of some fractures, which allows them to shear at large angles to the maximum stress. This gives further evidence that the friction coefficient of fractures at Soultz is not homogeneous but rather varies in a wide range (Figure 7).

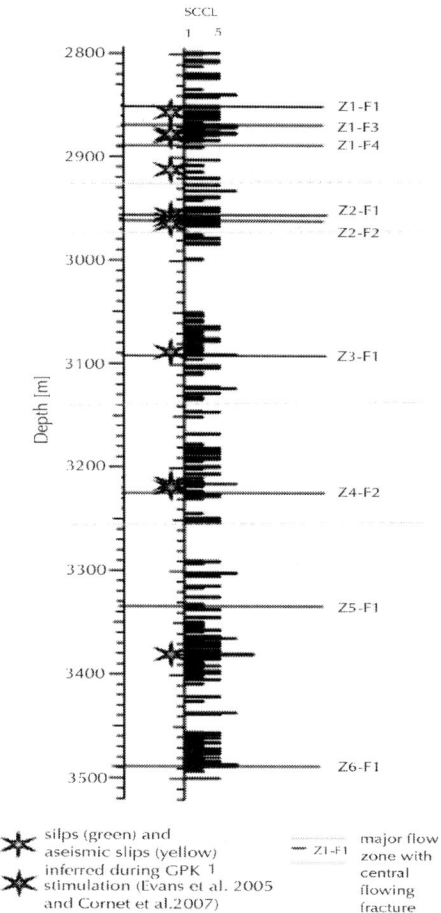

Figure 7: SCCL, major flow zones, and the occurrence of aseismic fault slip. In the well GPK1 between 2,800 and 3,500 m. SCCL and the location of (aseismic) slips during GPK1 stimulation have been identified by Cornet et al. ([1997]) and Evans et al. ([2005a]). Major flow zones are surrounded by zones of high SCCL. Slip is restricted to hydrothermally altered clay zones with high SCCL. The occurrence of aseismic slips is restricted to clay-rich flowing faults.

Aseismic movements on fractures have been directly observed in Soultz by Cornet et al. ([1997]). The SCCL of GPK1 indicates clay-rich intervals between 2,800 and 3,000, 3,050 and 3,100, 3,180 and 3,230, 3,340 and 3,410, and 3,450 and 3,500 m (Figure 7). The stars in this figure mark the shear movements, which have been induced during stimulation of GPK1, and which have been identified on image logs. All shear zones lay close to the flowing zones inside hydrothermally altered intervals, whereas most of the shear movements were aseismic (yellow stars). The higher number of aseismic movements at shallower depths is most probably related to the higher density of (large) fractures. A comparison between the orientation of these fractures and the orientation of S_H (Figure 8) reveals that some of the creeping fractures strike at an angle of >30° to S_H. If the Mohr-Coulomb failure criterion accounts for these fractures, their friction coefficients must be very low that shear is induced under the present conditions.

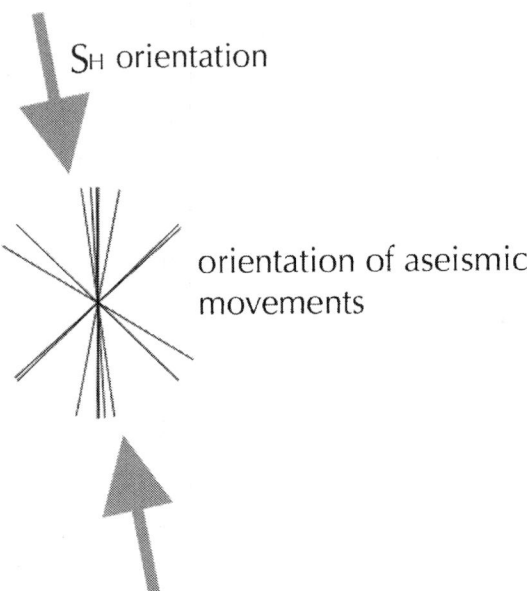

S$_H$ orientation

orientation of aseismic movements

Figure 8: Strike of the aseismic movements observed by Cornet et al. ([1997]) versus the orientation of SH. Some of the aseismic movements happened on fractures, which are oriented at an angle >30° to S_H, which indicates weakness of these fractures, as they are not favorably oriented for shear.

The correlation of aseismic movements with clay-rich intervals and their orientation at a significant angle to S_H supports the assumption of clay acting as some kind of lubricant on the fault zones. This makes these fractures prone for aseismic shearing, although they are not optimally oriented in the present stress field. Aseismic movements are assumed to take a big share of the movements induced during hydraulic stimulation, and some authors even assume that the major part of shearing happens aseismically (e.g., Schoenball et al. [2014]; Bourouis and Bernard [2007]). Further evidence for aseismic movements in Soultz in GPK1 (Bourouis and Bernard [2007]; Schmittbuhl et al. [2013]; Schmittbuhl et al. [2014]), GPK2 (Schoenball et al. [2014]; Calò et al. [2011]), and GPK3 (Calò et al.[2011]; Nami et al. [2008]) underlines the significance of clay on the structural reservoir evolution of the reservoir (Figure 8).

The SCCLs provide a unique opportunity to investigate the relation between seismic events and clay inside the reservoir. It is best constrained by calibration in the wells GPK1 and GPK3. In GPK2, the quality of the SCCL is bad due to the lack of logging data SCCL, and for GPK4, the location uncertainty of seismicity is too large (Gaucher, personal communication). Therefore, the analysis is focused on GPK1 and GPK3. In the well GPK1, the location and magnitude for several 19,000 seismic events have been determined (Jones et al. [1995]), and for GPK3, 22,000 events have been located (Dyer et al. [2003]; Dorbath et al. [2009]). As the SCCLs can only indicate clay in the proximity of the wells, the events in a radius of 100 m around the borehole are selected. For the remaining 5,600 events (4,200 for GPK1 and 1,400 for GPK3), the respective SCCL value of the depth, where they occurred, is determined. Then, the magnitude of the respective event is plotted against the SCCL value (Figure 9). The exciting result of this plot is that with increasing clay content, the maximum magnitude of seismic events is decreasing. Recent observations of Schorlemmer et al. ([2005]), Langenbruch and Shapiro ([2014]), and others suggest that low differential stress in weak zones prevents large seismic events, and it has long been assumed that the occurrence of large events in Soultz is restricted to fresh, i.e., unaltered granite, but it could never be directly observed. This figure suggests that the b-value of Soultz, although not constant in time, is also affected by the occurrence of hydrothermally altered zones. The b-value of the entire 2,000 stimulation test of GPK2 was determined to be 1.29 by Cuenot et al. ([2008]), and for the GPK3 stimulation, Dorbath et al. ([2009]) determined a b-value of 0.94. It would be interesting to compare the total clay content in both wells in order to find a correlation between the total clay content and the b-value. Unfortunately, the SCCL of the lower part of GPK2 could not be properly created due to missing image logs. The rather large value of 1.29 for GPK3 might be indicative of weak structures as such high values are normally only found in regions of crustal weakness (Amitrano [2003]).

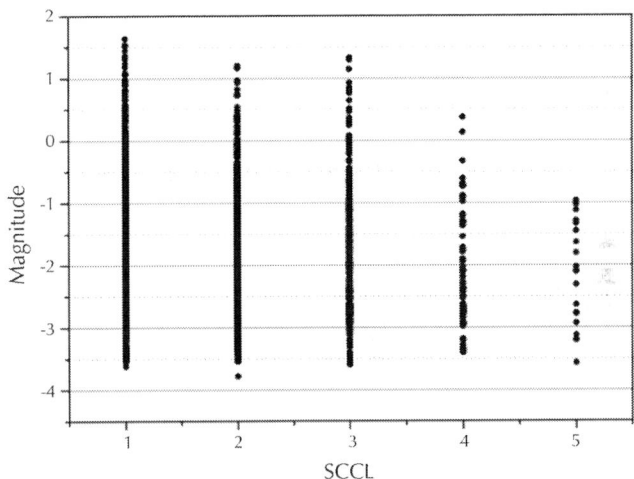

Figure 9: SCCL versus magnitude of induced seismic events during GPK1 and GPK3 stimulation. Seismic events with large magnitudes are restricted to low SCCL values, i.e., intervals with little clay inside fractures. With increasing clay content (higher SCCL), the maximum magnitudes of seismic events are smaller.

Therefore, the presence of large faults is most probably not the only reason for the different seismic behaviors of GPK2 and GPK3 as it was observed by Dorbath et al. ([2009]). The different b-values, which can be obtained from the seismic events induced during stimulation of these wells, could also be affected by the presence/absence of alteration zones (Figure 9).

There is evidence that hydrothermally altered zones not only affect the magnitude but also the evolution of seismicity. The seismic clouds of the GPK1 and GPK3 stimulation move downhole or uphole in steps, when the injection pressure is increased (Figure 10). Each pressure increase is marked by colored rectangles in Figure 10. The SCCLs provide a possible explanation of the reason of these steps. Each of these steps starts with a clay-rich zone and ends with an interval with little clay. As mentioned before, clay-rich intervals do not support large stresses and could act as small stress-decoupling horizons as observed by Cornet and Roeckel ([2012]) in the Paris Basin. They might therefore prevent the occurrence of seismic events or shear movements, or the events in such zones are very low-magnitude events, thus being too small to be measured, i.e., those slips are aseismic. With each pressure increase, the pressure front penetrates the decoupling horizon without inducing seismicity and the seismic events start just below or above this zone, migrating upward (as in GPK3, Figure 10b) or downward (as in GPK1, Figure 10a) until the next clay-rich zone is reached. This interval will then be overcome by the next pressure increase and so on, leading to a stepwise migration of microseismicity (Figure 10).

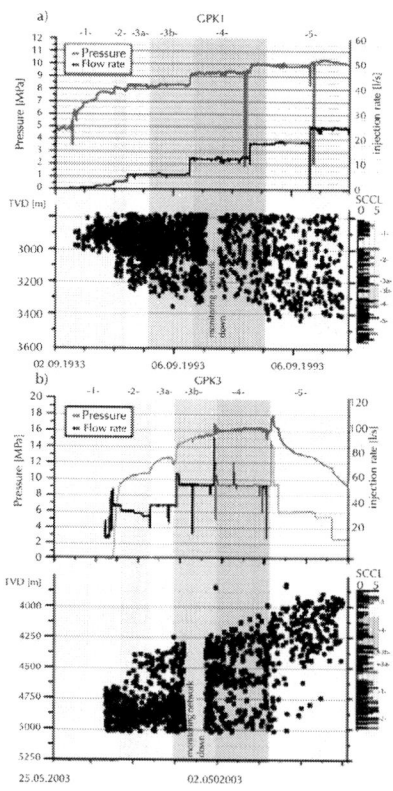

Figure 10: Evolution of seismicity in a radius of 100 m around the bore-holes during stimulation and SCCL. The upper part of the diagrams show the

stimulation pressure and below, the time-evolution of induced seismicity is shown versus depth. The lower part of the diagrams shows the SCCL for the respective depth intervals. Different sections are marked by numbers and colors indicating the pressure steps of injection. With increasing stimulation pressure, seismicity migrates downwards for (a)GPK1 and upwards for (b) GPK3. The steps might be caused by clay-rich intervals, which rather promote creep than seismic shearing. After each pressure increase, seismicity begins at depths, where the clay-content is low until the pressure front reaches a clay-rich interval. This interval is overcome during the next pressure increase, and seismicity starts just after the clay-rich zone migrating towards the next clay-rich zone, and so on. From step 5 on, we assume that the pressure front migrated too far away from the borehole so that a correlation with the (two-dimensional) SCCL is not possible any more.

CONCLUSIONS

The present SCCL method is an important basis to localize clay-rich zones as target zones for hydraulic stimulation and to identify fractures as candidates for aseismic movements. In order to optimally use the properties of hydrothermally altered zones, further effort has to be done on understanding of the processes affecting the geomechanical behavior of a geothermal reservoir. Once such processes are understood, it might become possible to exploit the properties of altered zones in order to increase the reservoir performance, while mitigating perceptible seismicity.

The occurrence of hydrothermally altered zones inside a geothermal reservoir can have large effects on many physical aspects, which are important for the performance of a geothermal system, and especially those related to induced seismicity. The observations at Soultz-sous-Forêts revealed that hydrothermal alteration lowers the mechanical strength of the Soultz granite and its fractures, which results in an inhomogeneously distributed friction coefficient. Geological units with low mechanical strength promote the occurrence of breakouts and can rotate the stress field as much as 90° from the mean orientation, which is indicated by high breakout-densities in clay-rich intervals and a deviation of their mean orientation.

A major result of this study is that hydrothermally altered zones can act as decoupling horizons, which change the local stress regime and thus significantly affect the seismicity induced during hydraulic stimulation at Soultz. It has been shown that large seismic events are restricted to fresh granite, whereas only small seismic events occur in clay-rich intervals. While this behavior has often been observed on the crustal scale, the present study for the first time confirms this effect on the scale of a geothermal reservoir.

Due to their low frictional strength and increased pore pressures, hydrothermally altered zones represent major target zones for hydraulic stimulation. In the future, EGS projects need to be structured in that prevention of large seismic events becomes a major achievement. Future stimulations could foster the creation of aseismic instead of seismic slip to increase the reservoir permeability, which requires knowledge on the location of such zones and advanced research towards the evolution of aseismic movements.

AUTHORS¿ CONTRIBUTIONS

CM interpreted the SCCL logs on the basis of borehole breakout analyses by Sahara et al. [2014] and of the thesis of Valley [2000] and on borehole analyses mainly conducted by Evans [2005], Evans et al. [2005]a; [2005]b] and Cornet et al. [1997]. Basis for the analysis were the seismic catalogues of GPK1 Jones et al. [1995] and GPK3 Dorbath et al. [2009]. CM wrote the manuscript and TK conducted the final revision. All authors read and approved the final manuscript.

ACKNOWLEDGEMENTS

This research was conducted within the portfolio topic GEOENERGIE of the Helmholtz Association of German Research Centres and was funded by Energie Baden-Wuerttemberg (EnBW), Germany. Thanks are given to GEIE Exploitation minière de la chaleur for providing the Soultz borehole data.

REFERENCES

1.   Akayuli C, Ofosu B, Nyako SO, Kwabena OO (2013) The influence of observed clay content on shear strength and compressibility of residual sandy soils. Int J Eng Res Appl 3(4):2538-2542

2. Amelung F, King G (1997) Earthquake scaling laws for creeping and non-creeping faults. Geophys Res Lett 24(5):507-510 doi:10.1029/97gl00287

3. Amitrano D (2003) Brittle-ductile transition and associated seismicity: experimental and numerical studies and relationship with the b value. J Geophys Res Solid Earth 108(B1):2044 doi:10.1029/2001jb000680

4. Babcock EA (1978) Measurement of subsurface fractures from dipmeter logs. AAPG Bull 62(7):15

5. Bachmann CE, Wiemer S, Goertz-Allmann BP, Mena B, Catalli F (2012) Why geothermal energy research needs statistical seismology. In: Thirty-seventh workshop on geothermal reservoir engineering. Stanford University, Stanford, California. p 8

6. Bartier D, Ledésert B, Clauer N, Meunier A, Liewig N, Morvan G, Addad A (2008) Hydrothermal alteration of the Soultz-sous-Forêts granite (hot fractured rock geothermal exchanger) into a tosudite and illite assemblage. Eur J Mineral 20:131-142doi:10.1127/0935-1221/2008/0020-1787

7. Boness NL, Zoback MD (2006) A multiscale study of the mechanisms controlling shear velocity anisotropy in the San Andreas Fault Observatory at Depth. Geophysics 71(5):F131-F146 doi:10.1190/1.2231107

8. Bourouis S, Bernard P (2007) Evidence for coupled seismic and aseismic fault slip during water injection in the geothermal site of Soultz (France), and implications for seismogenic transients. Geophys J Int 169(2):723-732 doi:10.1111/j.1365-246X.2006.03325.x

9. Brudy M, Zoback MD, Fuchs K, Rummel F, Baumgartner J (1997) Estimation of the complete stress tensor to 8 km depth in the KTB scientific drill holes: implications for crustal strength. J Geophys Res Solid Earth 102(B8):18453-18475 doi:10.1029/96jb02942

10. Brune JN (1968) Seismic moment, seismicity, and rate of slip along major fault zones. J Geophys Res 73(2):777-784 doi:10.1029/JB073i002p00777.

11. Byerlee JD (1978) Friction of rocks. Pure Appl Geophys 116(4):615-626 doi:10.1007/bf00876528.

12. Calò M, Dorbath C, Cornet FH, Cuenot N (2011) Large-scale aseismic motion identified through 4-D P-wave tomography. Geophys J Int 186(3):1295-1314 doi:10.1111/j.1365-246X.2011.05108.

13. Chang S-H, Avouac J-P, Barbot S, Lee J-C (2013) Spatially variable fault friction derived from dynamic modeling of aseismic afterslip due to the 2004 Parkfield earthquake. J Geophys Res Solid Earth 118(7):3431-3447 doi:10.1002/jgrb.50231

14. Charléty J, Cuenot N, Dorbath L, Dorbath C, Haessler H, Frogneux M (2007) Large earthquakes during hydraulic stimulations at the geothermal site of Soultz-sous-Forêts. Int J Rock Mechanics Mining Sci 44(8):1091-1105 doi:10.1016/j.ijrmms.2007.06.003

15. Chéry J, Zoback MD, Hickman S (2004) A mechanical model of the San Andreas fault and SAFOD Pilot Hole stress measurements. Geophys Res Lett 31(15):L15S13 doi: 10.1029/2004gl019521

16. Cornet FH (2012) The relationship between seismic and aseismic motions induced by forced fluid injections. Hydrogeol J 20(8):1463-1466 doi:10.1007/s10040-012-0901-z

17. Cornet FH, Roeckel T (2012) Vertical stress profiles and the significance of ¿stress decoupling¿. Tectonophysics 581:13 doi:10.1016/j.tecto.2012.01.020

18. Cornet FH, Helm J, Pointrenaud H, Etchecopar A (1997) Seismic and aseismic slips induced by large-scale fluid injections. Pure Appl Geophys 150(3):563-58 doi:10.1007/s000240050093

19. Cornet FH, Bérard T, Bourouis S (2007) How close to failure is a granite rock mass at a 5 km depth? Int J Rock Mechanics Mining Sci 44(1):47-66 doi:10.1016/j.ijrmms.2006.04.008

20. Crawford BR, Faulkner DR, Rutter EH (2008) Strength, porosity, and permeability development during hydrostatic and shear loading of synthetic quartz-clay fault gouge. J Geophys Res Solid Earth 113(B3): doi:10.1029/2006jb004634

21. Cuenot N, Dorbath C, Dorbath L (2008) Analysis of the microseismicity induced by fluid injections at the EGS site of Soultz-sous-Forêts (Alsace, France): implications for the characterization of the geothermal reservoir properties. Pure Appl Geophys 165(5):797-828 doi:10.1007/s00024-008-0335-7

22. Dezayes C, Genter A, Valley B (2010) Structure of the low permeable naturally fractured geothermal reservoir at Soultz. Cr Geosci 342(7¿8):517-530 doi:10.1016/j.crte.2009.10.002

23. Dieterich JH (1978) Time-dependent friction and the mechanics of stick¿slip. Pure Appl Geophys 116(4¿5):790-806 doi:10.1007/bf00876539

24. Dolan JF, Sieh K, Rockwell TK, Yeats RS, Shaw J, Suppe J, Huftile GJ, Gath EM (1995) Prospects for larger or more frequent earthquakes in the Los Angeles metropolitan region. Science 267(5195):199-205 doi:10.1126/science.267.5195.199

25. Dorbath L, Cuenot N, Genter A, Frogneux M (2009) Seismic response of the fractured and faulted granite of Soultz-sous-Forêts (France) to 5 km deep massive water injections. Geophys J Int 177(2):653-675 doi:10.1111/j.1365-246X.2009.04030.x

26. Dyer BC, Baria R, Michelet S (2003) Soultz GPK3 stimulation and GPK3-GPK2 circulation May to July 2003 seismic monitoring report. Semore Seismic report, GEIE.

27. Economides MJ, Nolte KG, Ahmed U (1989) Reservoir stimulation. Prentice Hall, Michigan.

28. Evans KF, Genter A, Sausse J (2005a) Permeability creation and damage due to massive fluid injections into granite at 3.5 km at Soultz: 1. Borehole observations. J Geophys Res-Sol Ea 110(B04203):19 doi:10.1029/2004jb003168

29. Evans KF, Moriya H, Niitsuma H, Jones RH, Phillips WS, Genter A, Sausse J, Jung R, Baria R (2005) Microseismicity and permeability enhancement of hydrogeologic structures during massive fluid injections into granite at 3 km depth at the Soultz HDR site. Geophys J Int 160(1):389-412 doi:10.1111/j.1365-246X.2004.02474.x

30. Fabriol H, Beauce A, Genter A (1994) Jones R (1994) induced microseismicity and its relation with natural fractures - the HDR example of Soultz (France)

31. Faulkner DR, Jackson CAL, Lunn RJ, Schlische RW, Shipton ZK, Wibberley CAJ, Withjack MO (2010) A review of recent developments concerning the structure, mechanics and fluid flow properties of fault zones. J Struct Geol 32(11):1557-1575 doi:10.1016/j.jsg.2010.06.009

32.   Genter A, Traineau H (1992) Hydrothermally altered and fractured granite as an HDR reservoir in the EPS-1 borehole, Alsace, France. In: Seventeenth workshop on geothermal reservoir engineering. Stanford University, Stanford, California. p 6

33.   Genter A, Traineau H (1996) Analysis of macroscopic fractures in granite in the HDR geothermal well EPS-1, Soultz-sous-Forêts, France. J Volcanol Geoth Res 72(1¿2):121-141 doi:10.1016/0377-0273(95)00070-4

34.   Genter A, Evans K, Cuenot N, Fritsch D, Sanjuan B (2010) Contribution of the exploration of deep crystalline fractured reservoir of Soultz to the knowledge of enhanced geothermal systems (EGS). Cr Geosci 342(7¿8):502-516 doi:10.1016/j.crte.2010.01.006

35.   Groos J, Zeiß J, Grund M, Ritter J (2013) Microseismicity at two geothermal power plants in Landau and Insheim in the Upper Rhine Graben, Germany. EGU General Assembly, Vienna.

36.   Gutenberg B, Richter C (1954) Seismicity of the earth and associated phenomena. Princeton University Press, Princeton.

37.   Häring MO, Schanz U, Ladner F, Dyer BC (2008) Characterisation of the Basel 1 enhanced geothermal system. Geothermics 37(5):469-495 doi:10.1016/j.geothermics.2008.06.002

38.   Heinicke J, Fischer T, Gaupp R, Götze J, Koch U, Konietzky H, Stanek K-P (2009) Hydrothermal alteration as a trigger mechanism for earthquake swarms: the Vogtland/NW Bohemia region as a case study. Geophys J Int 178(1):1-13 doi:10.1111/j.1365-246X.2009.04138.x

39.   Holmes RR, Jones LM, Eidenshink JC, Godt JW, Kirby SH, Love JJ, Neal CA, Plant NG, Plunkett ML, Weaver CS, Wein A, Perry SC (2013) U.S. Geological Survey natural hazards science strategy - promoting the safety, security, and economic well-being of the nation. US Geological Survey Circular 1383-F:79

40.   Ikari MJ, Saffer DM, Marone C (2009) Frictional and hydrologic properties of clay-rich fault gouge. J Geophys Res Solid Earth 114(B5): doi:10.1029/2008jb006089

41.   Ikari MJ, Marone C, Saffer DM (2011) On the relation between fault strength and frictional stability. Geology 39(1):83-86 doi:10.1130/g31416.1

42. Jones RH, Beauce A, Jupe A, Fabriol H, Dyer BC (1995) Imaging induced microseismicity during the 1993 injection tests at Soultz-Sous-Forêts, France. World Geothermal Congress, Florence, Italy.

43. Kohli AH, Zoback MD (2013) Frictional properties of shale reservoir rocks. J Geophys Res Solid Earth 118(9):5109-5125 doi:10.1002/jgrb.50346

44. Kohonen T (1984) Self-organization and associative memory. Springer series in information sciences, vol 8. Springer, Berlin.

45. Langenbruch C, Shapiro SA (2014) Gutenberg-Richter relation originates from Coulomb stress fluctuations caused by elastic rock heterogeneity. J Geophys Res Solid Earth 119(B2):15 doi:10.1002/2013jb010282

46. Ledésert B, Hebert R, Genter A, Bartier D, Clauer N, Grall C (2010) Fractures, hydrothermal alterations and permeability in the Soultz Enhanced Geothermal System. Cr Geosci 342(7¿8):607-615 doi:10.1016/j.crte.2009.09.011

47. Majer EL, Baria R, Stark M, Oates S, Bommer J, Smith B, Asanuma H (2007) Induced seismicity associated with Enhanced Geothermal Systems. Geothermics 36(3):185-222 doi:http://dx.doi.org/10.1016/j.geothermics.2007.03.003

48. Meixner J, Schill E, Gaucher E, Kohl T (2014) Inferring the in situ stress regime in deep sediments: an example from the Bruchsal geothermal site. Geothermal Energy 2(1):1-17doi:10.1186/s40517-014-0007-z

49. Meller C, Genter A, Kohl T (2014) The application of a neural network to map clay zones in crystalline rock. Geophys J Int 196(2):837-849 doi:10.1093/gji/ggt423

50. Meller C, Kontny A, Kohl T (2014b) Identification and characterization of hydrothermally altered zones in granite by combining synthetic clay content logs with magnetic mineralogical investigations of drilled rock cuttings. Geophys J Int 20: doi:10.1093/gji/ggu278

51. Meunier A (2005) Clays. Springer, Berlin.

52. Moore DE, Lockner DA (2007) Friction of the smectite clay montmorillonite: a review and interpretation of data. In: Dixon T (ed) The seismogenic zone of subduction thrust faults, Columbia Univ. Press, New York. pp 317-345

53.   Morrow C, Radney B, Byerlee J (1992) Chapter 3 frictional strength and the effective pressure law of montmorillonite and illite clays. In: Brian E, Teng-fong W (eds) International geophysics, vol Volume 51, Academic, 88. pp 69-88 doi:10.1016/S0074-6142(08)62815-6

54.   Mulargia F, Castellaro S, Ciccotti M (2004) Earthquakes as three stage processes. Geophys J Int 158(1):98-108 doi:10.1111/j.1365-246X.2004.02262.x

55.   Nami P, Schellschmidt R, Schindler M, Tischner T Chemical Stimulation Operations for Reservoir Development of the Deep Crystalline HDR/EGS System at Soultz-sous-Forêts (France) In: Thirty-Second Workshop on Geothermal Reservoir Engineering, Stanford, California, January 28¿30, 2008 2008. Stanford University, p 11

56.   Ruina A (1983) Slip instability and state variable friction laws. J Geophys Res Solid Earth 88(B12):10359-10370 doi:10.1029/JB088iB12p10359

57.   Rummel F, Klee G (1995) State of stress at the European HDR candidate sites Urach and Soultz

58.   Sahara D, Schoenball M, Kohl T, Mueller B (2014) Impact of fracture networks on borehole breakout heterogeneities in crystalline rock. Int J Rock Mech Mining Sci 71:301-309doi:10.1016/j.ijrmms.2014.07.001

59.   Sausse J (2002) Hydromechanical properties and alteration of natural fracture surfaces in the Soultz granite (Bas-Rhin, France). Tectonophysics 348(1¿3):169-185doi:10.1016/s0040-1951(01)00255-4

60.   Schleicher AM (2005) Clay mineral formation and fluid-rock interaction in fractured crystalline rocks of the Rhine Rift System: case studies from the Soultz-Sous-Forêts granite (France) and the Schauenburg Fault (Germany). Ruprecht-Karls-Universität, Heidelberg, Inaugural Dissertation.

61.   Schleicher AM, Van der Pluijm BA, Solum JB, Warr LN (2006) Origin and significance of clay-coated fractures in mudrock fragments of the SAFOD borehole (Parkfield, California). Geophys Res Lett 33(L16313):5doi:10.1029/2006GL026505

62. Schmittbuhl J, Lengliné O, Zaepfel , Cornet FH, Cuenot N (2013) Genter a seismic and aseismic slip in EGS reservoir: an experimental approach. European Geothermal Congress, Pisa, Italy.

63. Schmittbuhl J, Lengliné O, Cornet F, Cuenot N, Genter A (2014) Induced seismicity in EGS reservoir: the creep route. Geothermal Energy

64. Schoenball M, Baujard C, Kohl T, Dorbath L (2012) The role of triggering by static stress transfer during geothermal reservoir stimulation. J Geophys Res Solid Earth 117(B9):doi:10.1029/20 12jb009304

65. Schoenball M, Dorbath L, Gaucher E, Wellmann JF, Kohl T (2014) Change of stress regime during geothermal reservoir stimulation. Geophys Res Lett 41(4):1163-1170doi:10.1002/2013gl058514

66. Scholz CH (2010) The mechanics of earthquakes and faulting. Cambridge University Press, Cambridge.

67. Schorlemmer D, Wiemer S (2005) Earth science microseismicity data forecast rupture area. Nature 434(7037):1086-1086doi:10. 1038/4341086a

68. Schorlemmer D, Wiemer S, Wyss M (2005) Variations in earthquake-size distribution across different stress regimes. Nature 437(7058):539-542doi:10.1038/nature04094

69. Sibson RH (2009) Rupturing in overpressured crust during compressional inversion - the case from NE Honshu, Japan. Tectonophysics 473(3¿4):404-416doi:http://dx.doi.org/10.1016/j. tecto.2009.03.016

70. Smith SAF, Holdsworth RE, Collettini C, Imber J (2007) Using footwall structures to constrain the evolution of low-angle normal faults. J Geol Soc 164(6):1187-1191doi:10.1144/0016-76492007-009

71. Takahashi M, Mizoguchi K, Kitamura K, Masuda K (2007) Effects of clay content on the frictional strength and fluid transport property of faults. J Geophys Res Solid Earth 112(B8):doi:10.102 9/2006jb004678

72. Tembe S, Lockner DA, Wong T-F (2010) Effect of clay content and mineralogy on frictional sliding behavior of simulated gouges: binary and ternary mixtures of quartz, illite, and montmorillonite. J Geophys Res Solid Earth 115(B3):doi:10.1029/2009jb006383

73. Valley B (2007) The relation between natural fracturing and stress heterogeneities in deep-seated crystalline rocks at Soultz-sous-Forêts (France), dissertation. ETH Zürich, Zürich.

74. Valley B, Evans K (2000) Stress estimates from analysis of breakouts and drilling-induced tension fractures in GPK1 and GPK4. Synthetic final report, vol EC contract ENK5-2000-00301. ETH, Zurich.

75. Valley B, Evans K (2003) Strength and elastic properties of the Soultz granite. In: Zürich E (ed) Synthetic 2nd year report, Zürich, Switzerland, 2003. vol EC Contract SES6-CT-2003-502706, ETH, Zürich. p 6

76. Valley B, Evans KF (2007) Stress state at Soultz-sous-Forêts to 5 km depth from wellbore failure and hydraulic observations. 32nd workshop on geothermal reservoir engineering, Stanford.

77. Valley B, Evans KF (2010) Stress heterogeneity in the granite of the Soultz EGS reservoir inferred from analysis of wellbore failure. World Geothermal Congress, Bali, Indonesia.

78. (1995) Origin and mineralogy of clays. Springer, Heidelberg.

79. Voisin C, Cotton F, Di Carli S (2004) A unified model for dynamic and static stress triggering of aftershocks, antishocks, remote seismicity, creep events, and multisegmented rupture. J Geophys Res Solid Earth 109(B6): doi:10.1029/2003jb002886

80. Wu FT (1978) Mineralogy and physical nature of clay gouge. Pure Appl Geophys 116(4¡5):655-689 doi:10.1007/bf00876531

81. Wu FT, Blatter L, Roberson H (1975) Clay gouges in the San Andreas Fault System and their possible implications. Pure Appl Geophys 113(1):87-95 doi:10.1007/bf01592901

82. Zoback MD, Harjes H-P (1997) Injection-induced earthquakes and crustal stress at 9 km depth at the KTB deep drilling site, Germany. J Geophys Res 102(B8):18477-18491 doi:10.1029/96jb02814

83. Zoback MD, Kohli A, Das I, McClure M (2012) The importance of slow slip of faults during hydraulic-fracturing stimulation of shale gas reservoirs.

Study on Compatibility of Polymer Hydrodynamic Size and Pore Throat Size for Honggang Reservoir

Dan-Dan Yin[1, 2], Yi-Qiang Li[1, 2], Bingchun Chen[3], Hui Zhang[3], Bin Liu[3], Qingle Chang[1, 2], and Yanyue Li1, [2]

[1]EOR Research Institute, China University of Petroleum, Beijing 102249, China

[2]Key Laboratory of Petroleum Engineering, China University of Petroleum, Beijing 102249, China

[3]Exploration and Development Research Institute of Jilin Oilfield, PetroChina, Songyuan 138000, China

ABSTRACT

Long core flow experiment was conducted to study problems like excessive injection pressure and effective lag of oil wells during the polymer flooding in Honggang reservoir in Jilin oilfield. According to the changes in viscosity and hydrodynamic dimensions before and after polymer solution was injected into porous media, the compatibility of polymer hydrodynamic dimension and the pore throat size was studied in this experiment. On the basis of the median of radius R of pore throats in rocks with different permeability, dynamic light scattering method (DLS) was adopted to measure the hydrodynamic size Rh of polymer solution with different molecular weights. The results state that three kinds of 1500mg/L concentration polymer solution with 2000×10^4, 1500×10^4, and 1000×10^4 molecular weight matched well with the pore throat in rocks with permeability of 300mD, 180mD, and 75mD in sequence. In this case, the ratios of core pore throat radius median to the size of polymer molecular clew R/R_h are 6.16, 5.74, and 6.04. For Honggang oil reservoir in Jilin, when that ratio ranges from 5.5 to 6.0, the compatibility of polymer and the pore structure will be relatively better.

INTRODUCTION

Polymer flooding can improve the water-oil mobility ratio in the oil layer, effectively enlarge the swept volume, and be capable of enhancing oil recovery up to 10%, and this technology has already been carried out in Daqing oilfield [1–4]. The polymers used for flooding (partially hydrolyzed polyacrylamide (HPAM)) are trackless coils consisting of one or more twined molecular chains in the solution [5], and the clew size is affected by many factors, such as the relative molecular weight of polymer, polymer concentration, and the ionic concentration of the sample water [6–8]. Theoretically, for some reservoirs, under a certain concentration, with increasing relative molecular weight of the polymer, the viscosity improves, the hydrodynamic size increases, and the control ability over fluidity gets stronger [9, 10]. However, during the real injection process, the polymer used as oil-displacing agent will be naturally selected by the pore throat sizes as flowing through porous medium instead of entering the pores and throats in the porous mediums completely [11]. When the hydrodynamic size is way larger than the throat size of rock and under normal injection pressure, the polymer molecules will accumulate in the pipeline or at the core entrance and only a small amount of relative small molecular clews can enter the core [12]. Even though under some external forces, molecular clews with larger size may manage to enter the cores, the molecular structure will be damaged, and the displacement will be ineffective [13]; in addition, the migration distances for the polymer molecules entering the cores will be relatively short, which also make it difficult to generate effective displacement ability. Taking effect inconspicuously in oil wells is a typical issue at the mine field, and some objective intervals will also have some issues like pressure skyrocketing and failing to inhale, both of which will weaken the inject ability. So how to improve the effect of polymer flooding and further enlarge the swept volume so as to enhance the reserve utilization degree has been a research hotspot [14]. Therefore, it is of paramount importance for formulating the scheme and developing polymer flooding technology to study the matching relation between polymer and reservoirs with different permeability through analyzing the compatibility of hydrodynamic size and pore throat size. Polymer solution flooding experiment was conducted with 10cm long core by predecessors [11], and the compatibility of polymer hydrodynamic dimension and the pore throat size was studied just according to the core inlet pressure. However, long cores were not used to test pressure distribution in the core yet. Besides, the changes of viscosity and the molecular size of the effluent were not tested in other experiments.

In this paper, long cores with different permeability and 4 kinds of polymer with different molecular weight in X Block of Honggang reservoir located in Jilin oilfield were taken as the samples. The core flow test was used to detect the propagation rule of pressure during polymer injection process and judge the migration features of polymer. The changes in the viscosity of polymer solution and the clew sizes pre- and postcore injection were tested, so as to study the compatibility of hydrodynamic size and pore throat size.

EXPERIMENT

Experimental Materials and Apparatus

The polymers utilized were 4 kinds of HPAM named after p1, p2, p3, and p4 in this paper with different relative molecular weights of 800×10^4, 1500×10^4, 2000×10^4, and 2500×10^4, respectively, and the effective solid content was 80%. The water used for preparing liquid was taken from the injection water on site of Honggang reservoir in Jilin oilfield with 1033.4mg/L salinity. The ion composition and concentration are exhibited in Table 1. Water for preparing the fluid was prepared by filtering sample water on 0.22μm membranes and storing. The formation fracture pressure of target Block X is 52MPa.

Table 1: Water quality parameters of the on-site injection water from Honggang oilfield in Jilin

Ion	K+, Na+	Mg2+	Ca2+	Cl–	SO_4^{2-}	HCO_3^-
Concentration (mg/L)	162.5	36.7	87.8	127.6	164.6	454.2

The homogeneous long core used in the experiment was entirely pressed by outcrops with different radiuses. First, the outcrops were crushed into particles, and then particles were pressed into the cylindrical core with inorganic cementing agent under certain pressure. Their pore sizes are close to each other according to the mercury injection experiment. Compared with the quartz sand used for making normal core, the particle shape and surface properties of the outcrop sand are more similar to a real core because they have the same mineral composition. The core had 2.5cm radius and 100cm length. According to the reservoir permeability distribution of X Block in Jilin Honggang oilfield, representative cores were chosen with 300mD, 180mD, and 75mD air permeability, 148mD, 87mD, and 30mD water permeability, and 23.4%, 22.11%, and 20.32% porosity, and their pore throat radius medians were 3.14μm, 2.13μm, and 1.51μm, respectively, according to the mercury injection test.

Main experimental facilities are as follows: self-designed long core physical simulation equipment (Figure 1)—including a long core holding unit with surveying points (the surveying points in the middle divided the model into two equal parts), incubator, pressure sensor with high precision, data acquisition system, ISCO pump with constant velocity and pressure, HAAKE Rotational Rheometer, and microburette.

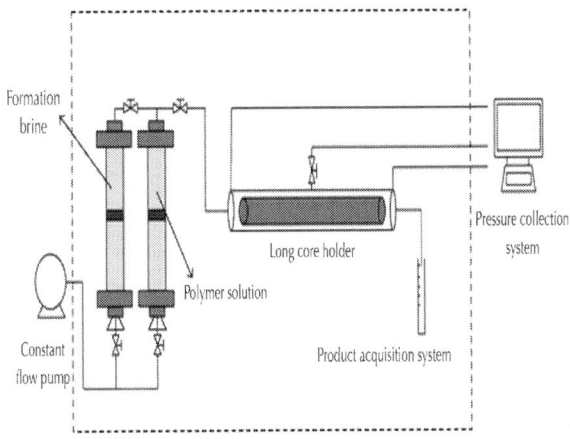

Figure 1: Flow chart of injection experiment.

Experimental Method

Prepare the polymers with different molecular weights into a mother solution with 5000mg/L concentration by using the on-site injection water and leave it to stand for 4h and then dilute it into 1500mg/L target solution. After that, shift 5 of warring blender shear apparatus was used to simulate the shearing action of the shot hole for 10s before storage.

Determination of Viscosity

The viscosities of the polymer solutions before and after injection were determined at a shear rate of $7.34s^{-1}$ with HAAKE Rotational Rheometer-6000 at 55°C.

Determination of Polymer Molecular Clew Size

American Brookhaven BI-200SM wide angle dynamic/static light scattering with 523nm wavelength was adopted, and the concentration of polymer solution was 100mg/L. DLS method has wide measurement range, high measurement speed, and less sample amount.

Mobility Test

- Vacuumize the core and saturate the formation water, measure the core permeability by water, and calculate the porosity.
- Conduct water flooding and polymer flooding under 55°C, record the injection pressure, collect the produced liquid, and measure relevant parameters.
- Conduct water flooding under 55°C for a second time, record the injection pressure, and measure relevant parameters.

Referring to the responding time of polymer flooding in Daqing oilfield, the linear velocity of effective flow for the fluid flowing in the stratum was figured out to be 1 m/d. Therefore, 3–5PV polymer solution was injected continuously with an injection flow of 0.1mL/min until a stabilized situation was achieved.

EXPERIMENTAL RESULTS AND DISCUSSION

Injection Characters of Polymer Solution

The propagation performance of polymers propagating in the porous medium was evaluated primarily, that is to say, whether there is retention phenomenon during the injection process. The propagation rules of pressure were evaluated principally. If the polymer solution has preferable propagation performance in the porous medium, then no block will emerge and the pressure will drop equally throughout homogeneous cores. Moreover, the resistance coefficients of the front half and latter half of the core will be close. Resistance coefficient (RF) is the ratio between water and polymer solution mobility:

$$RF = \frac{\lambda_w}{\lambda_p} = \frac{K_w/\mu_w}{K_p/\mu_p}.$$

(1)

In the equation, RF is the resistance coefficient (a nondimensional parameter); λ_w is the water mobility; λ_p is the polymer solution mobility; K_w is the water phase permeability, in mD; K_p is the polymer solution permeability, in mD; μ_w is the viscosity of water, in mPa·s; μ_p is the viscosity of polymer solution, in mPa · s.

When the hydrodynamic size of the polymer is larger than the throat size of rock, the polymer molecules will accumulate at the core entrance or travel short distances in the core, so the pressure and resistance coefficients in the front and back long core sections are very different. When the hydrodynamic size of the polymer has a very good compatibility with the throat size of rock, the polymer molecules will travel smoothly in the core and are distributed evenly throughout the core. So the pressure and resistance coefficients in the front and back long core sections are much closed. Figure 2 shows the pressure variation with the change in injection volume when 4 kinds of polymer solutions were injected into 3 kinds of cores with different permeability. A surveying point on the experimental model divided the model into two parts (PX-1 and PX-2 are the pressure at the core entrance and the middle part, resp.). When the pressure on the surveying points goes up, it indicates that the polymer solution has migrated to this part. During the process of chemical agent injection, it is the pressure around the entrance that rose first, and the pressure at the middle increased along with the chemical agent migration. Four kinds of polymer solution with different molecular weights were injected into cores with 300mD, 180mD, and 75mD permeability, and their variation tendencies are shown in Figures 2(a), 2(b), and 2(c), respectively.

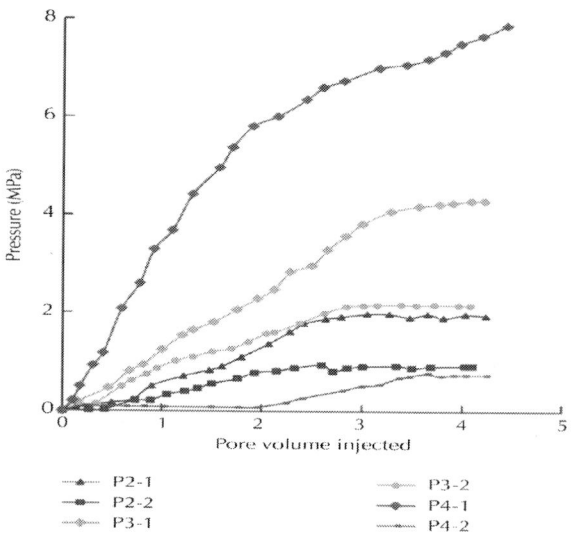

(a)

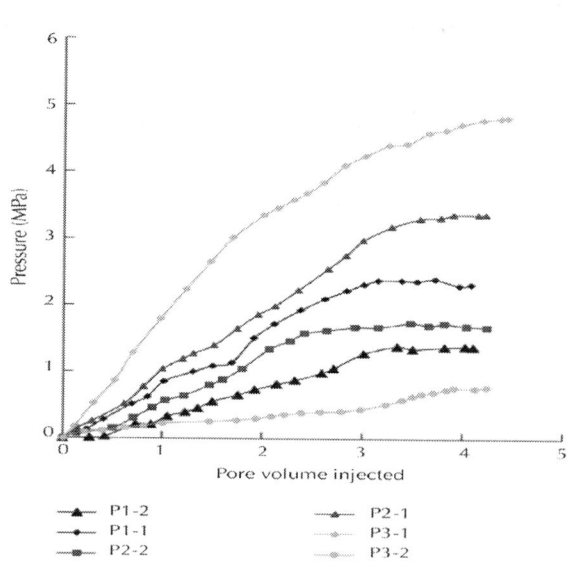

(b)

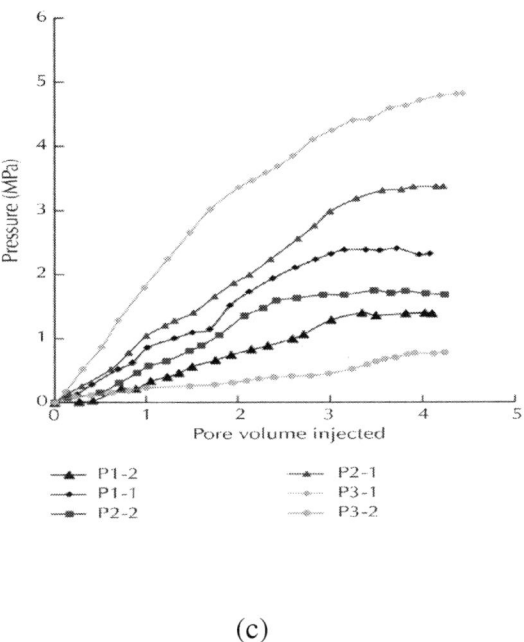

(c)

Figure 2: Injection pressures variation with injection volume of different polymer solutions ((a) Kg: 300mD; (b) Kg: 180mD; (c) Kg: 75mD).

Figure 2(a) shows the pressure variation with injection volume by injecting solutions prepared by p2, p3, and p4 polymers into a core with 300mD permeability. As shown in the figure, when a small amount of p4 polymer solution (25 million molecular weight) was injected, the pressure at the entrance surged while pressure in the middle showed no obvious variation, the differential pressure between the front and back sections was 7.12MPa, and the resistance coefficients of the two sections were 389.3 and 3.7, respectively. This meant that the polymer solution mainly accumulated at the core entrance, and only when the injection volume reached 2.0PV, pressure in the middle would finally increase slightly. The injected polymer molecules also accumulated on the front part of the core and failed to achieve remote migration. The phenomenon in return explained on-site injection well pressure build up and injection failure issues at the oilfield. Pressure unable to delivery efficiently will lead to injection pressure build up or even fracture at the stratum. Moreover, during the injection process of p4 solution, the pressure was always changing, which meant that the polymer molecular clews were oversized and mismatched with the size of formation pores. The displacement agent of p3 solution was prepared by polymers with 20 million molecular weight, the pressure increased gradually from the core entrance (P3-1) to the middle part (P3-2) with a steady rising velocity, the differential pressure between the entrance and the middle part was 2.13MPa, and the resistance coefficients of the front and back sections were 106.5 and 105, respectively. This reflected the steady migration process of chemical agent in the model and preferable compatibility of polymer molecular clew size and the pore size. During the process of p2 solution injection, the pressure increased gradually from the core exit (P2-1) to the entrance (P2-2) with steady rising velocity, which further reflects the stable migration process of chemical slug in the model. However, due to the low injection pressure in the injection system, the differential pressure was 1.03MPa, and the resistance coefficients of the front and back parts were 53 and 46, respectively. The excessive velocity of the injection system in the large pores led to a failure of controlling the fluidity, which means that polymer molecular clew was undersized and mismatched with the pore size.

Figure 2(b) is a diagram of the pressure variation with the injection volume as injecting polymers p1, p2, and p3 into a core with 180mD permeability. The pressure of solution prepared by p3 increased at the entrance (P3-1), while there was no change in the middle (P3-2), the differential pressure was 3.3MPa, and the resistance coefficients of the front and back sections were 201 and 39, respectively. This means that the polymer solution mainly accumulated at the core entrance, and the oversized polymer molecular clew in the solution prepared by p3 mismatched with the pore size in a stratum with 180mD permeability. The pressure variation tendency of the solution prepared by p2 shows was approximately linear increase from the core entrance to the middle part, the differential pressure between the surveying points at the entrance and in the middle was 1.69MPa, and the resistance coefficients of the front and back sections were 84.5 and 83, respectively. This means that the chemical agent migrated steadily in the core and implies a good compatibility of polymer molecular clew size and pore size. Figure 2(c) shows the pressure variation with the injection volume as p1, p2, and p3 polymers were injected into a core with 75mD permeability. The pressure variation of solution prepared by p1 increased uniformly from the core entrance to the middle part, the differential pressure between the entrance and the middle part was 2.56MPa, and the resistance coefficients of the front and back sections were 128 and 122, respectively, which illustrated that the polymer molecular clew size matched well with the pore size.

Changes in Polymer Solution Viscosity

Table 2 shows the viscosity changes in produced liquid after 4 kinds of polymer solution were injected into 3 kinds of cores with different permeability.

Table 2: Viscosity and its loss rate in the produced liquid with 4 kinds

Kg, mD	p1		p2		p3		p4	
	Viscosity, mPa·s	Loss rate, %	Viscosity, mPa·s	Loss rate, %	Viscosity, mPa·s	Loss rate, %	Viscosity, mPa·s	Loss rate, %
300	12.36	4.31	21.13	5.95	41.95	8.26	51.94	31.12
180	11.96	7.45	20.27	9.77	33.23	27.32	34.89	53.73
75	11.77	9.16	13.76	38.77	22.85	50.03	21.48	71.51

The viscosities of the four kinds of polymer solution with 1500mg/L concentration were 75.40mPa·s, 45.73mPa·s, 22.47mPa·s, and 12.92mPa·s in sequence before injection.

The polymer viscosity loss was mainly due to the shearing and absorption effect, and several conclusions are suggested by the data: for cores with the same permeability, the viscosity loss rate of the produced liquid increased with the polymer molecular weight. This means that with increasing molecular weight, the length of polymer molecular chain after hydration increased and the polymer became easier to be broken under stress during the migration process in the pores, leading to an increase in the viscosity loss rate. Under the same polymer molecular weight conditions, smaller core permeability means smaller pore size, more complex structure, which will result in stronger shearing effect on polymer, larger specific surface area of core, more severe absorption effect on polymers, and, as a consequence, a serious viscosity loss of the polymer solution.

Combined with the flow test in the cores and conducting some further analysis, p1, p2, and p3 matched well with the cores with

300mD, 180mD, and 75mD permeability, and these three kinds of polymer solutions had little viscosity loss rate in the cores with corresponding permeability and stable pressure. To further verify their matching relations, microanalysis is required to measure changes in hydrodynamic dimensions.

Changes in the Polymer Hydrodynamic Dimensions

Four kinds of polymer solution with different molecular weight and 1500mg/L concentration were injected into cores with 300mD, 180mD, and 75mD permeability at 0.1mL/min velocity until the pressure was stable, after which the hydrodynamic dimensions of produced liquid at the exit were measured, and both the polymer solution and produced liquid were diluted to 100mg/L. The polymer molecular clew sizes and light scattering distribution results are reported in Table 3.

Table 3: Changes in the hydrodynamic dimensions of the produced liquid from polymer solution

Kg mD	p1		p2		p3		p4	
	h, m	Loss rate, %	h, m	Loss rate, %	h, m	Loss rate, %	h, m	Loss rate, %
300	0.24	4.0	0.34	5.6	0.46	9.8	0.52	18.8
180	0.23	8.0	0.32	11.1	0.42	17.6	0.45	29.7
75	0.22	12.0	0.28	22.2	0.34	33.3	0.36	43.8

The hydrodynamic dimensions of 4 kinds of polymer solutions p1, p2, p3, and p4 with 100mg/L concentration are 0.25μm, 0.36 μm, 0.51 μm, and 0.64 μm, respectively (the order was ranked by their molecular weights).

As shown in Figure 3, the average hydrodynamic dimension R_h of the polymer solution with 800 × 10^4, 1500 × 10^4, 2000×10^4, and 2500×10^4 molecular weights before injection were 0.25 μm, 0.36 μm, 0.51 μm, and 0.64 μm, respectively, and the peak values are relatively concentrated. After flowing over the core, the R_h of polymer solution decreased, and, for the same core permeability conditions, a larger polymer molecular weight means greater drop in hydrodynamic dimension; for solutions with the same polymer molecular weight, lower permeability means greater drop degree of the hydrodynamic dimension, which is mainly due to the low permeability and small pore size. This makes a molecular clew with relatively larger size unable to pass through the pores with its molecular structure undamaged. After passing through cores with different permeability, the peak value of polymer R_h moved towards left and the proportion of polymers with relatively smaller hydrodynamic dimensions increased. In contrast, the proportion of polymers with relatively larger R_h decreased and the phenomenon can be explained in this way: when the polymers passed through the pores and throats in the core, the quantity of polymer with larger R_h decreased with decreasing core permeability, and that was consistent with the conclusion about viscosity loss rate.

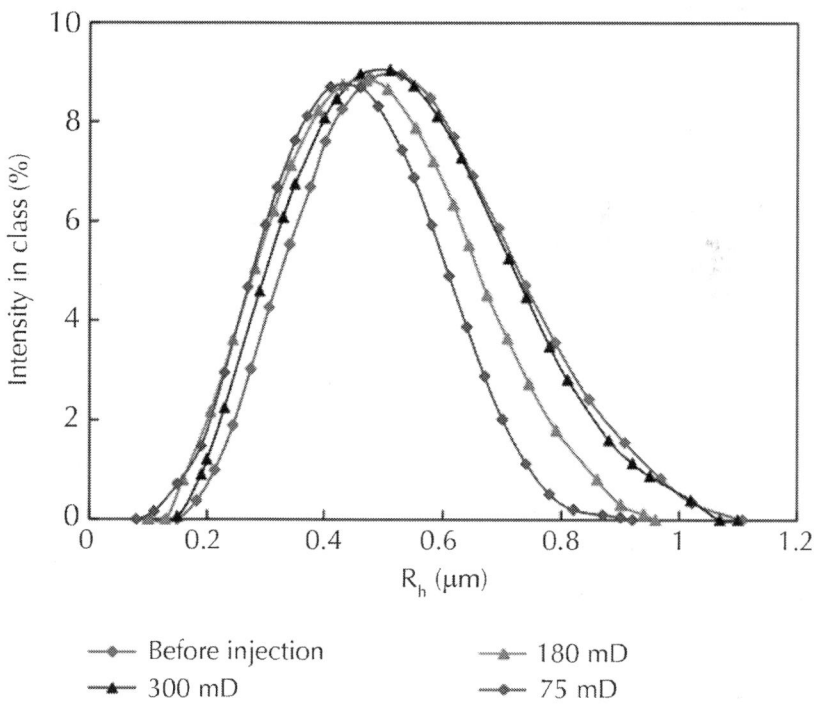

(a)

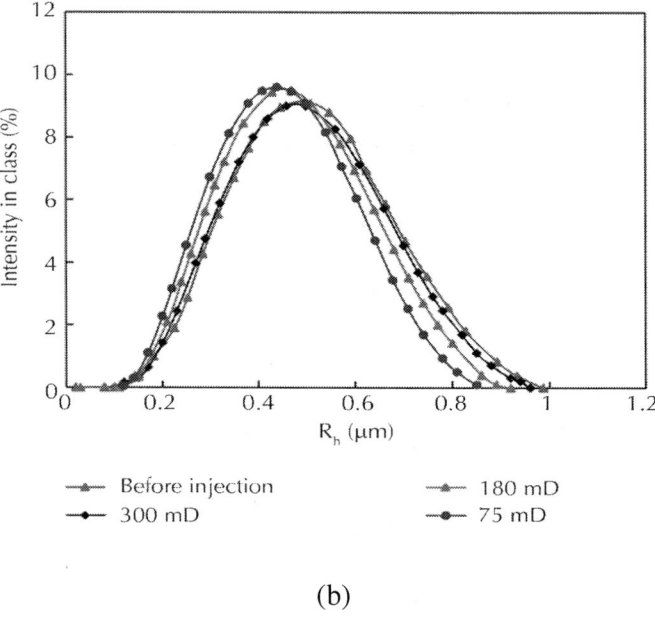

(b)

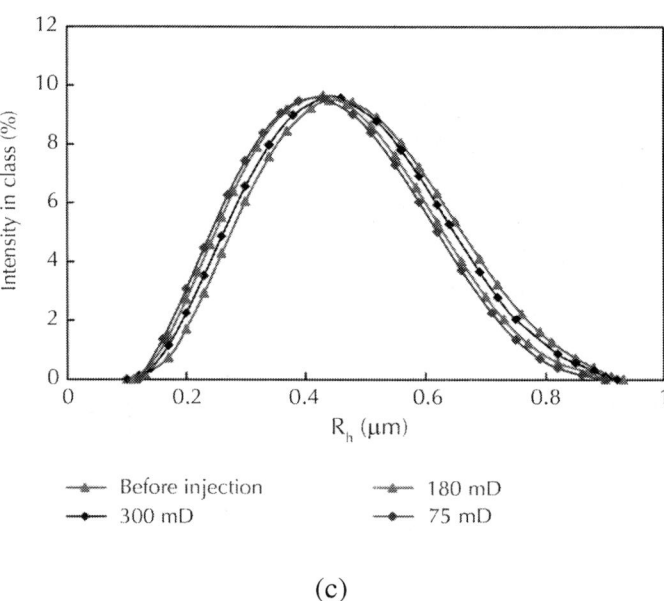

(c)

Figure 3: R_h distribution of polymer solution with different relative molecular mass before and after injection of different permeability cores.

Compatibility of Polymer Molecular Clew Size and Pore Throat Size

As different researchers defined differently the pore radius, the water used for preparing polymer solution had different salinity, and, for different reservoir conditions, the specific values R/R_h for representative polymer and core compatibility were hard to determine.

In this paper, the reservoir conditions in Honggang oilfield were used, and the polymer hydrodynamic size and the pore throat radius were unified at their average values. Table 4shows the relevant data.

Table 4: Calculation results of the specific value of pore throat radius median to average hydrodynamic dimension

Polymer parameters		Core parameters		R/Rh
M	Rh/ m	Kg/mD	R/ m	
2000 × 104	0.51	300	3.14	6.16
1500 × 104	0.38	180	2.18	5.74
1000 × 104	0.25	75	1.51	6.04

To avoid polymer retention in the pore throats in the stratum of Honggang reservoir, the core permeability, polymer concentration, relative molecular weight, and the chemical composition of the solvent water were taken into consideration. Generally, R/R_h ranges from 5.5 to 6.0. For stratum with 300mD, 180mD, and 75mD permeability, polymers with relative molecular weight lower than 2000×10^4, 1500×10^4, and 1000×10^4 are recommended, and if the shearing and degradation effect during preparation and injection processes are taken into consideration, the relative molecular weights of polymers can still be increased modestly.

CONCLUSIONS

- When the polymer solution was injected into cores with different permeability, the injection pressure had varying degrees of increase. Three kinds of polymers with 800×10^4, 1500×10^4, and 2000×10^4 molecular weights matched well with the pore throats in cores with permeability of 300mD, 180mD, and 75mD, respectively. They showed preferable inject ability and no block at the injection end and propagated uniformly in the cores.

- For Honggang reservoir in Jilin oilfield, when the specific value R/R_h (pore throat radius median/polymer clew size) is among 5.5–6.0, the matching relation between polymer and the pore structure of cores is preferable and can better suit the pore throat structure in the reservoirs.

- During the practical application of polymer flooding at Honggang oilfield in Jilin, for stratum with 300mD, 180mD, and 75mD permeability, polymers with 2000×10^4, 1500×10^4, and 1000×10^4 molecular weights were recommended, and if the shearing and degradation effect during preparation and injection processes are taken into consideration, the relative molecular weights of polymers can still be increased modestly.

ACKNOWLEDGMENTS

The authors would like to express their appreciation for the financial support received from National Natural Science Foundation (51074172) and EOR Institute of China University of Petroleum for permission to publish this paper.

REFERENCES

1.   D.-M. Wang, J.-C. Cheng, J.-Z. Wu, and G. Wang, "Application of polymer flooding technology in Daqing Oilfield," Acta Petrolei Sinica, vol. 26, no. 1, pp. 74–78, 2005.

2.   C. Jiecheng, S. Xinguang, W. Yan, J. Luo, W. Lei, and X. Qing, "Cases studies on polymer flooding for poor reservoirs in daqing oilfield," in Proceedings of the Asia Pacific Oil and Gas Conference and Exhibition, 2007.

3.   N. Jin-gang, "Practices and understanding of polymer flooding enhanced oil recovery technique in Daqing oilfield," Petroleum Geology Oilfield Development in Daqing, vol. 5, article 024, 2004.

4.   D. Wang, H. Dong, C. Lv, X. Fu, and J. Nie, "Review of practical experience of polymer flooding at Daqing," SPE Reservoir Evaluation & Engineering, vol. 12, no. 3, pp. 470–476, 2009. ·
.

5.   L. Wenli, M. Desheng, N. Xiaobin et al., "Study on matching relation between polymer molecular size and pore size for conglomerate reservoir," in Proceedings of the International Petroleum Technology Conference Challenging Technology and Economic Limits to Meet the Global Energy Demand (IPTC ‹13), pp. 1778–1786, March 2013.

6.   P. Chen, Z. Shao, and Y. Liu, "Method for determining relative molecular mass of polymer in medium-low permeable reservoirs," Petroleum Geology & Oilfield Development in Daqing, vol. 3, article 040, 2005.

7.   M. Li, Z. Liu, X. Song, B. Ma, and W. Zhang, "Effect of metal ions on the viscosity of polyacrylamide solution and the mechanism of viscosity degradation," Journal of Fuel Chemistry and Technology, vol. 40, no. 1, pp. 43–47, 2012.

8.   Z. Xiutai, W. Zengbao, and Q. Guangmin, "Study on influence factors of the initial viscosity of HPAM solution," Chemical Engineering of Oil & Gas, vol. 3, article 15, 2009.

9.   J. Cheng, D. Wang, and J. Wu, "Molecular weight optimization for polymer flooding,"Acta Petrolei Sinica, vol. 21, pp. 102–106, 2000.

10.   S. G. Goodyear, J. D. Johnston, T. A. Lawless, and C. L. Woods, "Measurement of polymer retention levels representative of the formation," in Proceedings of the SPE/DOE Enhanced Oil Recovery Symposium, 1992.

11.   L. Wenli, M. Desheng, L. Meiqin, L. Gang, N. Xiaobin, and L. Qingxia, "Matching relation between HPAM polymer DQ3500 and pores of reservoir rock," Procedia Engineering, vol. 18, pp. 261–270, 2011.

12.   T. Lotsch, T. Muller, and G. Pusch, "The effect of inaccessible pore volume on polymer coreflood experiment," in Proceedings of the SPE Oilfield and Geothermal Chemistry Symposium, 1985.

13.   Q. He, T. Young, G. P. Willhite, and D. W. Green, "Measurement of molecular weight distribution of polyacrylamides in core

effluents," SPE Reservoir Engineering, vol. 5, no. 3, pp. 333–338, 1990.

14. D. G. Hatzignatiou, U. L. Norris, and A. Stavland, "Core-scale simulation of polymer flow through porous media," Journal of Petroleum Science and Engineering, vol. 108, pp. 137–150, 2013. · ·

Chapter 6

The Fate of Injected Water in Shale Formations

Hongtao Jia[1], John McLennan[1], and Milind Deo[1]

[1]Department of Chemical Engineering, University of Utah, Salt Lake City, UT, USA

ABSTRACT

It is well known that only about a third of water injected for hydraulic fracturing of shales is recovered. It is important to understand the fate of this injected water. The amount of water infiltrating the matrix is determined by a number of parameters such as the pressure differential between the fracture and the matrix, the capillary pressure relationships in the fractures and in the matrix and other petrophysical properties of the formation. In this paper, we provide a breakdown for the various possible water losses depending on the reservoir, fracture and operating parameters. A set of capillary pressure relationships for the formation were first created based on the basic mineralogy and the total organic carbon (TOC) content. Fracture capillary pressure also changed depending on the concentrations and types of proppants in the fractures. Two basic end members can be defined – silicistic and dolomitic with different amounts of TOC. The capillary pressure relationships ranged from oil wet, neutral to water wet. Different porosity and permeability combinations were also examined. Amounts of water relative to the total amount injected that would infiltrate the formation were compiled as the operating conditions (pressures) and formation properties changed. This calculation shows that the infiltration due to the various phenomena are not sufficient to account for the water losses if the formations are strongly oil wet. In addition, situations where water blockages occur due to these multiphase flow effects were identified and the loss of productivity due to this phenomenon was quantified both for gas and for oil production. The study was conducted using a discrete-fracture network simulator developed at the University of Utah. A realistic (non-orthogonal) representation of a complex fracture network was employed in the study. Realistic representation of distribution and retention of these aqueous fracturing fluids is essential for optimizing hydraulic fracturing treatment volumes.

INTRODUCTION

The growth in producing hydrocarbons from unconventional reservoirs (shales) has been phenomenal. The production of liquids from the Eagle Ford play grew to about 52 million barrels in 2011 [1] (Figure 1).

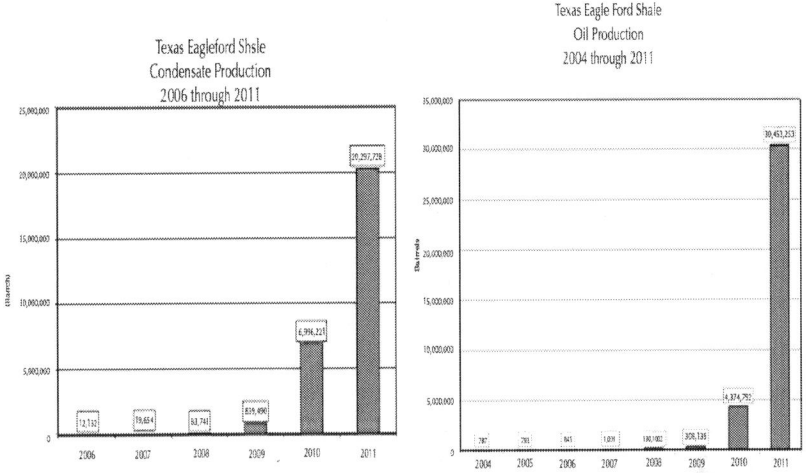

Figure 1: The phenomenal growth in production of liquids from shales with Eagle Ford. In just over a three-year period, insignificant production has been transformed to over 52 million barrels of liquids in 2011.

The growth in production is driven by improvements in hydraulic fracturing technology. Multistage fracturing using long horizontal wells is the common practice. Millions of gallons of water are pumped into the formation to create these fractures. Industry data reveals that only about a third of the injected water is typically recovered. The fate of injected water is of fundamental interest. Use of large quantities of water in fracturing has brought into question the sustainability of this type of completion and development practice. Furthermore, low water recovery has prompted environmental concerns about whether the injected water leaves the target formation with a potential of infiltrating and contaminating aquifers. The purpose of this paper was to examine the capability of the formation to imbibe the injected water based on different capillary pressure relationships.

TECHNICAL APPROACH

The Advanced Reactive Transport Simulator (ARTS) at the University of Utah was used to perform simulation studies (Figure 2). ARTS is a modular reservoir simulator that has been under development over a number of years [2-4]. The main idea of ARTS is to decouple the discretization methods from the physical models. The discretization methods in ARTS include the conventional finite difference, control-volume finite element and a generalized control volume method. These discretization methods could be coupled with a variety of physical models. The simplest physical model would be simulation of a single-phase gas with immovable water phase. Two-phase and three-phase black oil models are used to simulate primary production followed by water and polymer flooding. Thermal processes such as steam flooding, in-situ combustion, steam-assisted gravity drainage, etc. are represented in K-value based thermal-compositional models. In these models, the vapor-liquid equilibrium is calculated using the ratio between the vapor and the liquid phase composition of each component (K-value). ARTS also includes a geochemical module to simulate processes associated with carbon dioxide sequestration and reactions involving carbon dioxide, brine and rocks.

The use of a control volume finite element model as one of the discretization schemes allows multiphase simulation of complex reservoir geometries including a discrete fracture network representation of natural and hydraulic fractures.

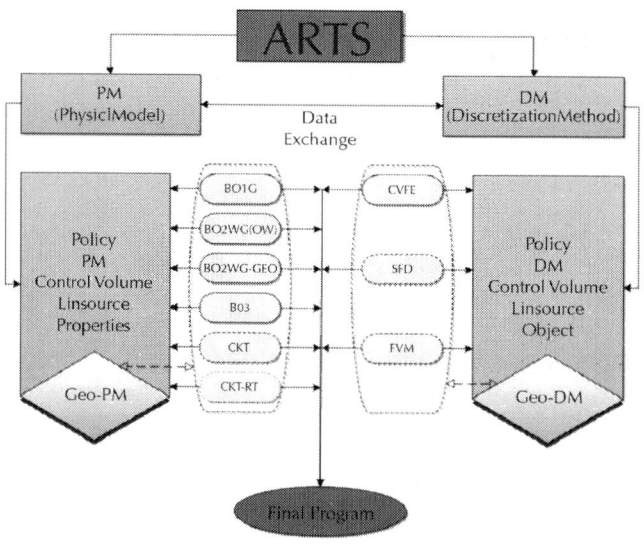

Figure 2: The framework used in simulating water injection and production in fractured systems. The discretization methods (DM) are decoupled from the physical models (PM).

We represented and simulated two different discrete fracture domains in this work – both with non-orthogonal features (Figure 3). It is common practice to represent and simulate hydraulic fractures as orthogonal features. However, it is evident that the fractures created are not perfectly perpendicular to the horizontal well. The microseimic cloud that is observed in a number of cases with multiple horizontal fractures (for example, [5]), shows fractures that are more complex than regularly spaced orthogonal features. It is true that there is no one to one correlation between the microseismic signatures and the shape and morphology of hydraulic fractures. However, there are a number of indications that point to the hydraulic fractures being more complex than simple orthogonal features.

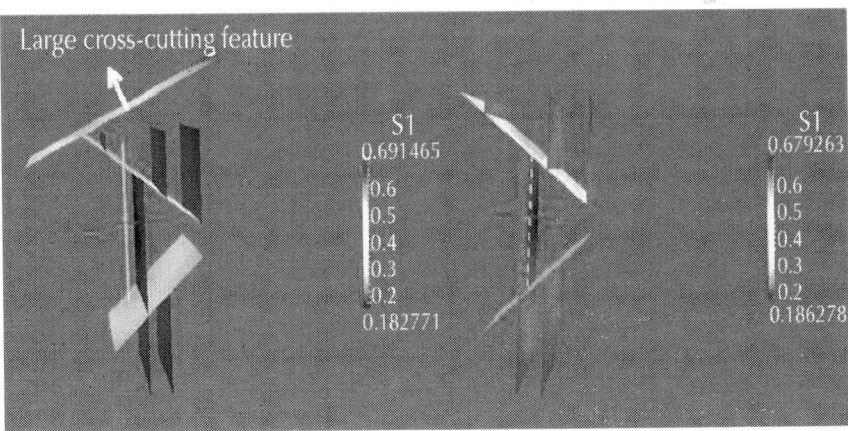

Figure 3: Figure showing two fractured systems simulated in this study.

The hydraulic fractures created interact with existing natural fractures. The role of natural fractures in production of fluids from shales is still an open question. The production behavior of both the gas and liquid reservoirs does not indicate a highly fractured system. On the other hand, when fracturing water is injected in a well, it is common to see interference in an adjacent well. This may be in the form of pressure interference or explicit breakthrough of water injected in the adjacent well. Pressure interference in and of itself does not indicate fluid transport to the well.

Capillary pressures for these shale reservoirs are not well characterized. The wettability of the reservoir rocks is also not well known. Al-Bazali et al. [6], measured sealing capacities of shale caprocks. This data provides some guidance for the capillary pressure values and relationships to use for these systems. The general capillary pressure relationship is given by:

Pc=2σcosθr

In this equation, P_c is the capillary pressure, σ is the interfacial tension between the immiscible fluids of interest, θ is the contact angle and r is the average pore radius. Al-Bazali et al.[6], were considering shales that were less than 10 nD in permeability. For the three shales studied, they measured entry pressures ranging from 470 psia to 750 psia. They calculated pore throat radii of about 30 nM for entry pressures of crude oil. For pore throats of less than 10 nM (Sondergeld et al. [7]), very large capillary pressures (two to three times those measured by Al-Bazali et al [7]) are possible.

There has been much discussion about wettability of shales. In this paper, we examined the differences in water recovery due to variations in wettability of the rock. The three sets of oil-water capillary pressures used in the study are shown in Figure 4.

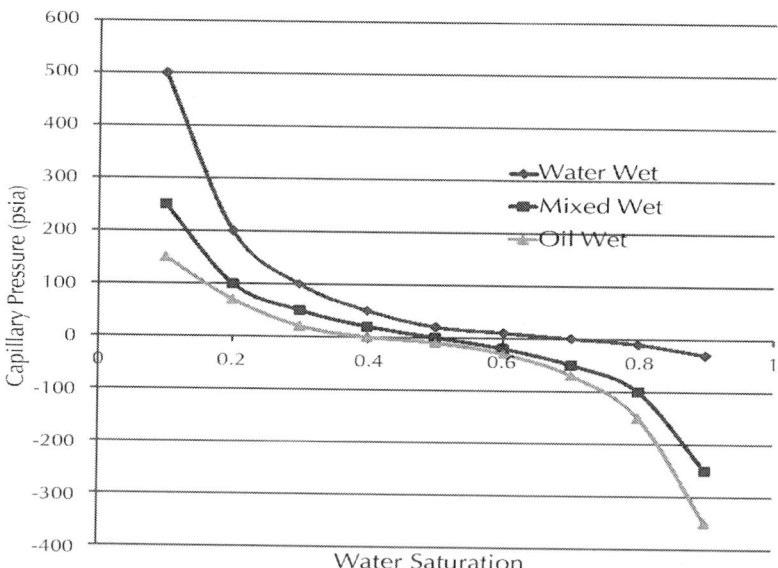

Figure 4: The three sets of capillary pressures used in this study.

Over most of the saturation range for the oil and mixed wet situations, the capillary pressures are negative, indicating a preference for oil as the wetting fluid. Other domain-specific parameters are shown in Table 1.

Table 1: Properties of the domain and simulations

Domain Size	260 feet X 260 feet X 100 feet
Initial Reservoir Pressure	2000 psia
Fracture Permeability	1000 mD
Porosity	20%
Matrix Permeability	0.5 mD
Water Injected	30000 barrels

Water recovery after one month (30 days) for each of the simulations was compiled. For the base case capillary pressures, the water recoveries for the three wetting scenarios and for the two domains (one with the cross-cutting fracture, and one without) are shown in Table 2.

Table 2: Water recoveries for the three wetting scenarios and for the two domains studied in this paper. Recoveries are for the base case where the initial reservoir pressure was 2000 psia and the matrix permeability was 0.5 mD

	Water Wet	Mixed Wet	Oil Wet
Water recovery ratio(With cross-cutting fracture)	21.53%	29.35%	36.28%
Water recovery ratio (Without the cross-cutting fracture)	22.97%	31.24%	38.39%

The water recoveries observed in the table above are consistent with water recoveries of about 20-40% listed in field observations. Water recoveries increase as we go from water wet to mixed wet to oil wet clearly indicating the tendency of the matrix to imbibe and hold water as the formation becomes more water wet. There is a 15% increase in water recovery as we go from water wet to the oil wet case. The presence of the long cross-cutting feature does not make a significant impact in recovery. The recovery does decrease as injected water is transported to longer distances – but the difference in recovery is only 1-2%.

In a number of shale reservoirs, the permeabilities are lower and the initial pressures are higher. To investigate the effects of these parameters on recovery, simulations were performed with 5000 psia initial pressure and 0.1 mD matrix permeability. Results of these simulations are tabulated in Table 3.

Table 3: Water recoveries for the three wetting scenarios and for the two domains studied in this paper. Recoveries are for the base case where the initial reservoir pressure was 5000 psia and the matrix permeability was 0.1 mD

	Water Wet	**Mixed Wet**	**Oil Wet**
Water recovery ratio(With cross-cutting fracture)	37.42%	40.17%	44.19%
Water recovery ratio (Without the cross-cutting fracture)	41.02%	44.61%	49.83%

Higher initial pressure results in higher water recoveries, particularly in the water wet cases. The differences between recoveries with and without the large cross-cutting feature are now between 4-5%. The differences between the different wettability cases however are reduced to only about 8% (compared to about 15%) as the largest difference the water wet and the oil wet scenarios.

At smaller pore radii, the capillary pressures are expected to be larger. One set of simulations were performed where the shape of the base case capillary pressures were maintained, but the capillary pressures were increased ten times for each of the saturation values. The resulting recoveries are tabulated in Table 4.

Table 4: Water recoveries for the three wetting scenarios and for the two domains studied in this paper. Recoveries are for the case where the capillary pressures were ten times the base case capillary pressures used. The shapes of the capillary pressure curves were the same as the ones used in Figure 4. The initial reservoir pressure was 5000 psia and the matrix permeability was 0.1 mD

	Water Wet	Mixed Wet	Oil Wet
Water recovery ratio(With cross-cutting fracture)	20.1%	27.15%	41.9%
Water recovery ratio (Without the cross-cutting fracture)	23.3%	30.2%	45.5%

As the capillary pressure increases, more water is retained. For mixed wet and oil wet scenarios, water saturation in the matrix area is lower (Figure 5). Similar relative difference between recoveries is maintained when recoveries are compared for domains with and without the large cross-cutting features. The system without the large cross-cutting fracture in this case returns on the average about 3% more water than when the large fracture exists. Water saturations for the domain without the large fracture are shown in Figure 6.

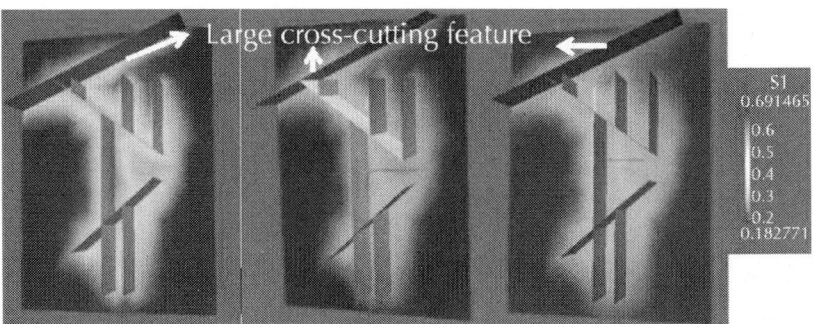

Figure 5: Figure showing water saturations in the matrix through one hydraulic fracture and interacting natural fractures. Left panel is for the water wet case, the middle panel is for the mixed wet case and the right panel is the oil wet case. As the wettability goes from water wet to oil wet the infiltration decreases increasing injected water recovery. In this particular example, the large cross-cutting feature does not take a significant amount of water off site.

CONCLUSIONS

Recovery of water injected for hydraulic fracturing in shales is only about 30%. There is a question of the fate of injected water. In this paper we studied water retention in shales for different shale wettability conditions. Two different domains where a hydraulic fracture intersected with a small existing network of natural fractures were used in the simulations. A specially developed framework that can handle representation of complex fracture networks was used for simulations. Capillary pressures in rocks containing very small pores tend to be high – of the order of 1000 psia. Three sets of capillary pressures – water wet, mixed wet and oil wet were examined. Simulations showed that a recovery of 20-30% is expected for typical water wet conditions, while a recovery of about 37%-48% is expected for oil wet scenarios. The recovery for mixed wet conditions fell between these two extremes. The recovery is reduced when a large cross-cutting fracture is introduced – but not significantly. That is because water will be recovered if the fractures are interconnected. Results discussed in this paper helped quantify the role of wettability in the recovery of water used for hydraulic fracturing. In this paper we assumed that the initial water saturation was low and that the water was immovable. If that is not the case, water saturation in the matrix and in the natural fractures, as well as the water-oil or water-gas relative permeability functions play significant roles in determining the water balance.

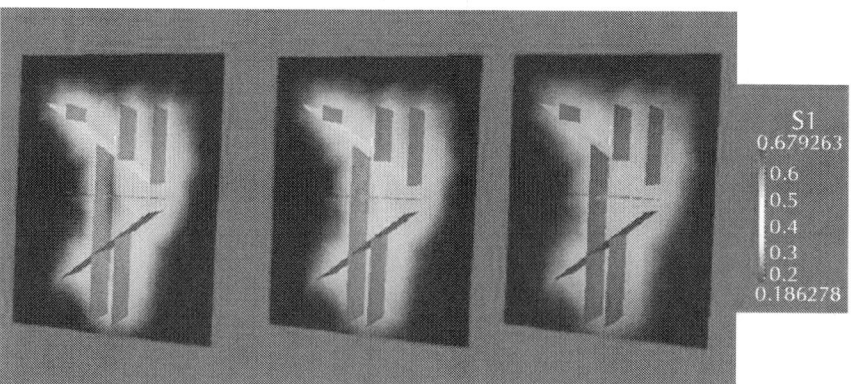

Figure 6: Figure showing water retained in the matrix through one hydraulic

fracture and interacting natural fractures. Domain without the large cross-cutting feature is used. Left panel is for the water wet case, the middle panel is for the mixed wet case and the right panel is the oil wet case. Water saturation scale is also shown. As the wettability goes from water wet to oil wet the infiltration decreases increasing injected water recovery.

REFERENCES

1.   Data from the Texas Railroad Commission- http://wwwrrc.state.tx.us/

2.   Y. K Yang, 2003Finite-Element Multiphase Flow Simulator, Ph.D. dissertation, University of Utah, 2003.

3.   Y Fu, 2007Multiphase Control Volume Finite Element Simulation of Fractured Reservoirs, Ph.D. dissertation, University of Utah, 2007.

4.   Z Gu, 2010A Geochemical Compositional Simulator for Modeling C O2 Sequestration in Geological Formations, Ph.D. dissertation, University of Utah, 2010.

5.   N. A Stegent, K Ferguson, and J Spencer, 2011Comparison of Frac Valves vs. Plug and Perf Completions in the Oil Segment of the Eagle Ford Shale: A Case Study, CSUG/SPE 148642, Paper presented at the Canadian Unconventional Resources Conference, Calgary, Canada, 1517November 2011.

6.   T. M Al-bazali, J Zhang, M. E Chenevert, and M. M Sharma, 2005Measurement of the Sealing Capacity of Shale Caprocks, SPE 96100, Paper presented at the Annual Technical Conference and Exhibition, Dallas, Texas, Oct 912

7.   M. E Curtis, R. J Ambrose, C. H Sondergeld, and C. S Rai, Structural Characterization of Gas Shales on the Micro- and Nano-scales, CSUG/SPE 137693, Paper presented at the Canadian Unconventional Resources and International Petroleum Conference held in Calgary, 1921October 2010

Microbial Enhanced Heavy Oil Recovery by the Aid of Inhabitant Spore-Forming Bacteria: An Insight Review

Biji Shibulal[1], Saif N. Al-Bahry[1],
Yahya M. Al-Wahaibi[2], Abdulkader E. Elshafie[1],
Ali S. Al-Bemani[2], and Sanket J. Joshi[1, 3]

[1]Department of Biology, College of Science, Sultan Qaboos University, 123 Muscat, Oman

[2]Petroleum and Chemical Engineering Department, College of Engineering, Sultan Qaboos University, 123 Muscat, Oman

[3]Central Analytical and Applied Research Unit, College of Science, Sultan Qaboos University, 123 Muscat, Oman

ABSTRACT

Crude oil is the major source of energy worldwide being exploited as a source of economy, including Oman. As the price of crude oil increases and crude oil reserves collapse, exploitation of oil resources in mature reservoirs is essential for meeting future energy demands. As conventional recovery methods currently used have become less efficient for the needs, there is a continuous demand of developing a new technology which helps in the upgradation of heavy crude oil. Microbial enhanced oil recovery (MEOR) is an important tertiary oil recovery method which is cost-effective and eco-friendly technology to drive the residual oil trapped in the reservoirs. The potential of microorganisms to degrade heavy crude oil to reduce viscosity is considered to be very effective in MEOR. Earlier studies of MEOR (1950s) were based on three broad areas: injection, dispersion, and propagation of microorganisms in petroleum reservoirs; selective degradation of oil components to improve flow characteristics; and production of metabolites by microorganisms and their effects. Since thermophilic spore-forming bacteria can thrive in very extreme conditions in oil reservoirs, they are the most suitable organisms for the purpose. This paper contains the review of work done with thermophilic spore-forming bacteria by different researchers.

BACKGROUND

Oil productions have been experiencing decline in many parts of the world due to the oil field maturity, and example of such includes the major oil fields in the North Sea [1]. Another major factor which causes downgrade is the increasing energy demands due to global population growth and the difficulty in discovering new oil fields as an alternative to the exploited oil fields. Therefore, there is an urge to find out alternative technologies to increase oil recovery from existing oil fields around the world. It is a fact that fossil fuels will still remain the key source of energy, regardless of the gross investments in other energy sources such as biofuels, solar energy, and wind energy. Current global energy production from fossil fuels represents about 80–90% with oil and gas typifying about 60% [2]. Cossé [3] stated that during the process of oil production, between 30 and 40% of oil can be contributed by primary oil recovery, while additional 15–25% can be recovered by secondary methods such as water injection leaving behind about 35–55% of oil as residual oil in the reservoirs. The focus of many enhanced oil recovery technologies is this residual oil, and it amounts to about 2–4 trillion barrels [4] or about 67% of the total oil reserves [5]. For many oil companies, residual oil recovery is at present unavoidable, and so there is a perpetual hunt for a cheap and efficient technology which will raise the global oil production as well as the productive life of many oil fields. The recovery of this residual oil is accomplished by enhanced oil recovery (EOR) or tertiary recovery methods which are used in oil industry to increase the production of crude oil. Most common tertiary recovery methods include chemical flooding, miscible CO_2 injection, and thermally enhanced oil recovery method which uses heat as a main source for the additional oil recovery [6]. Large quantities of residual oil in the depleted oil reservoirs could be regained by these EOR methods as the current primary and secondary extraction methods leave about two-thirds of the original oil in the reservoir. One of the potential EOR methods is microbial enhanced oil recovery (MEOR), which employs microorganisms to pull out the remaining oil from the reservoirs. Up to 50% of the residual oil can be extracted by this exceptionally low operating cost technology [7, 8]. The field trials of MEOR method project a chance to reverse the declining trend of oil production or at least to maintain a curve with a positive slope. This is achieved by the alteration of chemical and physical properties of reservoir rocks and crude oil by the microbial growth and metabolites produced [9]. MEOR can overcome the main hindrances of efficient oil recovery such as low reservoir permeability, high viscosity of the crude oil, and high oil-water interfacial tensions, which in turn result in high capillary forces retaining the oil within the reservoir rock [10].

THE REASONS FOR OIL TO GET LEFT BEHIND

The fundamental cause for leaving oil behind is economics. In general, the process of recovering oil from any conventional reservoir requires (a) a pathway which connects oil in the pore space of a reservoir to the surface and (b) sufficient energy in the reservoir to drive the oil to the surface. Lack of these requirements in the environment results in oil getting left behind. In this case, it is not economical to implement incremental development activities. In addition, all of the theoretically displaceable oil cannot be recovered, even if there is a pathway and adequate reservoir energy, due to the physics of fluid displacement in porous media [11].

ENHANCED OIL RECOVERY

The residual crude oil in reservoirs is up to 67% of the total petroleum reserves in the world, which in turn represents the relative inefficiency of the primary and secondary production techniques. Extraction of this trapped oil can be achieved by injecting chemicals (polymers or surfactants), gases (carbon dioxide, hydrocarbons, or nitrogen), or steam into the reservoir. The chemicals used for EOR must be compatible with the physical and chemical environments of oil reservoirs. The varying permeability of petroleum reservoirs is also a major concern in EOR processes. When water is injected to displace the oil, it preferentially flows through areas of highest permeability and bypasses much of the oil [12]. Thus, the conventional EOR methods to recover the entrapped crude oil seem not to be very efficient.

MICROBIAL ENHANCED OIL RECOVERY (MEOR)

MEOR is a tertiary oil recovery technique. Recovering oil usually requires three stages. At the primary recovery only 12% to 15% of the oil in the well is recovered without the need to introduce other substances into the well. The oil well is then flooded with water or other substances to drive out an additional oil (15% to 20%) from the well which is known as the secondary recovery. Tertiary recovery is the last phase which is accomplished through several different methods, including MEOR, for the additional extraction of trapped oil from the well. In principle, the process of MEOR results in some beneficial effects such as formation of stable oil-water emulsions reduced interfacial tension and clogging the high permeable zones. In in situ MEOR method, bacteria inoculated with water in to the well will progress into high-permeability zones at first. Then at a later stage they will grow and occlude those zones due to their size and the negative charge on their cell surface. This scenario helps to increase the sweep efficiency, and thus a more efficient oil recovery can be achieved [11, 13].

Microorganisms can synthesize useful products by fermenting low-cost substrates or raw materials. Therefore, MEOR can substitute chemical enhanced oil recovery (CEOR), which is a very pricey technology. In MEOR, the chosen microbial strains are used to synthesize compounds analogous to those used in CEOR processes which are very expensive, to increase the recovery of oil from depleted and marginal reservoirs. Furthermore, microbial products are biodegradable and have low toxicity [7, 14, 15]. Microbial technologies are becoming approved universally as lucrative and eco-friendly approaches to improve oil production [16, 17].

MEOR OUTCOMES

MEOR is based on two absolute justifications. Oil advancement through porous media is expedited by modifying the interfacial properties of the oil-water minerals. In such a system, microbial activity alters fluidity (viscosity reduction, miscible flooding); displacement efficiency (decrease of interfacial tension, increase of permeability); sweep efficiency (mobility control, selective plugging); and driving force (reservoir pressure).

The second principle is known as upgrading. In this case, the degradation of heavy oils into lighter ones occurs by microbial activity. Instead, it can also aid in the removal of sulphur from heavy oils as well as the removal of heavy metals.

Continuous research and successful applications affirm the fact that MEOR can be viewed as a potent technology [8, 22, 23] despite the existing disagreement by some groups [24]. However, successful MEOR field applications reported are specific for each well and published information to support economic advantages is lacking. MEOR is, therefore, considered as one of the promising future research areas with great preference as identified by the Oil and Gas in the 21st Century Task Force [24]. This is probably because MEOR is an alternate technology that may help in recovering the 377 billion barrels of oil that are unrecoverable by conventional technologies [8].

THE BYGONE DAYS OF MEOR

It was Beckman in 1926 [25] who suggested for the first time that microbes could be used to recover oil from porous media. Between 1926 and 1940, not many studies were held on this topic. In the 1940s, Zobell [26] started a series of systematic laboratory findings which marked the beginning of a new era of petroleum microbiology research with application in oil recovery. According to Zobell the main mechanisms behind oil release from porous media are processes such as bacterial metabolites that break up inorganic carbonates; bacterial gases which reduce the viscosity of oil, thereby increasing its flow; surface-active substances or wetting agents produced by some bacteria; and the high affinity of bacteria for solids to crowd off the oil films, processes by which bacterial products (gases, acids, solvents, surface-active agents, and cell biomass) releasing oil from the sand pack columns in wet labs were patented by Zobell. Later Updegraff et al. repeated [27, 28] Zobell's experiments and patented [29] the process which is based on the bacterial byproducts produced from cheap substrates like molasses to assist the oil recovery. The first field test was carried out in the Lisbon field, Union County, AR [30]. Kuznetsov et al. [31] concluded that anaerobic bacteria present in the oil deposits can utilize oil to form gaseous products (CH_4, H_2, CO_2, N_2). Kuznetsov's work demonstrated the technology of microbial flora activation of reservoirs, later advanced by Ivanov et al. [32]. Extensive research on MEOR was conducted in the 1960s and 1970s, in Czechoslovakia, Hungary, and Poland [33–35]. The field trials were based on the injection of mixed anaerobic or facultative anaerobic bacteria (Clostridium, Bacillus, Pseudomonas, Arthrobacterium, Micrococcus, Peptococcus, Mycobacterium, etc.) selected on their ability to generate gases, acids, solvents, polymers, surfactants, and cell biomass. At the same time, another technology named as selective plugging recovery has been recognized as an important additional mechanism for improving the oil recovery from water floods. This is achieved by producing polysaccharide slime in situ by an injected microbial system based on molasses. Microbes producing biopolymers of xanthan or scleroglucan types as viscosifying agents were isolated, which greatly enhanced oil recovery [36–38]. The investigations during 1970–2000 have demonstrated the basic nature and existence of indigenous microbiota in oil reservoirs, as well as reservoir characteristics essential to a successful MEOR application. It was also proved that the cyclic microbial recovery (single well stimulation), microbial flooding recovery, and selective plugging recovery are very effective. The technology based on activation of stratal microbiota was successfully developed in former Soviet Union [32, 39]. It can be concluded that the petroleum crisis during 1970s led to substantial MEOR research and later became a scientifically identified EOR method, supported by research projects carried out all over the world in countries such as the USA, Canada, Australia, China, Russia, Romania, Poland, Hungary, Czech Republic, Great Britain, Germany, Norway, and Bulgaria. Many international meetings were periodically organized on the MEOR topic with the publication of proceedings carrying the advances in the knowledge and practice of MEOR techniques. It is important to recognize and acknowledge the role of the U.S. Department of Energy (DOE), which sponsored MEOR basic research and field trials, as well as periodically organizing international meetings. Several books on MEOR were also published [40–42]. Grula et al. [43] developed a microbial screening method to isolate an anaerobic Clostridium species that produced gases, acids, alcohols, and surfactants. But all those strains isolated showed intolerance to high salt concentrations (>7%) which remained as a major problem. Success of in situ MEOR processes depends upon isolating microorganisms that can survive and produce the desired metabolic products in reservoirs containing hydrocarbons and saline water. Continuous investigations were done on different microbial species such as Clostridium species, Bacillus species, and Enterobacter for better adaptation to reservoir conditions. By the end of the 1990s, MEOR was recognized as a scientific and interdisciplinary technique for the increase of oil recovery.

In 1995, a survey of 322 MEOR projects in the USA showed that 81% of the projects successfully increased oil production, and neither of them had shown reduced oil production [7]. Today, MEOR technologies are well suited for application, when there is a need for oil crisis at a rate of 3 to 4%/year. Since 1980, the abolition of stripper wells has increased to 175% [9], and accordingly, within 15–25 years, the USA could have access to less than 25% of its remaining oil resources. MEOR technologies were very slowly recognized by industry even though a long history of MEOR activity exists, due to the lack of published data especially in widely available journals, as well as too little cooperation between microbiologists, reservoirs engineers, geologists, economists, and owner operators.

LABORATORY AND FIELD MEOR PROJECTS

Zobell [44] patented a process for the secondary oil recovery, using anaerobic, hydrocarbon-utilizing, and sulfate-reducing bacteria such as Desulfovibrio species in situ. He reported that the oil recovery mechanism was similar to Clostridium, where bacterial cells (and the hydrogenase enzyme system) produces the acids and ammonium hydroxide by using CO_2, water and nitrates present in the reservoir, which helps to enhance the release of oil from reservoir rock when supplied with nutrients.

Various "agroindustrial carbohydrates based" substrates are proposed as a suitable "carbon source" for MEOR applications, like molasses [17, 45]. Updegraff and Wren [27] proposed that fermentative bacteria such as Desulfovibrio use nutrients such as molasses to produce large amounts of organic acids and carbon dioxide to enhance oil recovery in wet labs. The process was patented by them in spite of the major drawback of Desulfovibrio species producing hydrogen sulfide which is not suitable for MEOR processes. MEOR research team at Sultan Qaboos University, Oman, have reported isolation, identification, and bioproducts production by spore-forming Bacilli spp., and its potential role in enhancing oil recovery at laboratory scale [13, 17–21]. Bond [46] injected 5,000 gal of agar medium containing sand and Desulfovibrio hydrocarbonoclasticus, which is no longer a valid species into a sandstone reservoir at a depth of 3,000 ft. The well initially produced 15 bbl/day. After the inoculum injection, the well was shut in for 3 months for the bacterial growth and action. The well, when it started the production again, produced 25 bbl/day.

Hitzman [47] patented a process of injecting bacterial spores along with nutrients into a reservoir. The spores would germinate in the reservoir and enhance oil recovery from reservoir rock. A medium containing molasses and spores of Clostridium roseum was passed through the sand-packed column saturated with oil and showed about 30% increase in release of oil.

Patents by Hitzman [36, 47] used microorganisms that utilized injected polymers and the byproducts of CO_2 floods, to produce products such as gases, acids, solvents, and surfactants for EOR.

In polymer floods, the injected organisms feed on polymer that is adsorbed on the reservoir rock. In CO_2 floods, the microbes feed on soluble compounds of carbon, nitrogen, and sulfur left behind by the CO_2-crude oil slug. The process was demonstrated in sand-pack, but no core or field tests are reported.

Knapp et al. [48] reported the isolation of 22 microorganisms that produce biopolymers and emulsifiers. Among them, one strain could thrive at 10% salt concentrations, over a pH range of 4.6 to 9.0, at temperatures up to 50°C, in presence of crude oil. They demonstrated that glucose, ammonium sulfate, and potassium phosphate were easily transported through sandstone cores. Viable bacterial cells in aqueous solutions of 2% NaCl and 0.01% $CaCl_2$ injected into these cores were not recovered in the effluent. The cores were inoculated with bacteria and nutrients such as glucose were added which resulted in a significant decrease in permeability. This could be because of the plugging of pores by the bacterial mass. The prominent bacteria indigenous to all of the cores treated were found as Pseudomonas sp., Bacillus sp., and Actinomycetes. A major problem in these experiments was the determination of the amount of plugging caused by injected bacteria and the amount by inhabitant ones. The problem existed even when cores were steam-sterilized and autoclaved. "Sterilization" of cores with chlorine dioxide helped to get rid of the problem, but the bacterial populations returned after 48 h incubation.

Johnson [49] studied 150 stripper wells in the USA that produced, on an average, 2 bbl/day, with no well head pressure. The reservoir porosities were 10 to 30%, depths 200 to 1,000 ft, with an average reservoir temperature of 38°C. In his study, he inoculated a mixed culture of Bacillus and Clostridium spp. (1 to 10 gal) with crude molasses and mineral salts as nutrients. Approximately 10 to 14 days were needed for the optimal growth of cells in the treated area of the reservoir. The results varied, but an average of 20 to 30% additional oil-in-place was recovered.

The preliminary field tests done by Petrogen, Inc., during 1977–88 in 24 wells with varying depths from 300 to 4,600 ft., demonstrated a pressure increase of 10 to 200 psi in 75% of the wells. Four wells doubled production for 6 months, and 12 increased production by 50% for 3 months. The average production increase was indicated as 42%; however, the final results remain to be reported [50]. Jack et al.

[51] considered that emulsification of viscous crude oil in situ is not a feasible method for EOR since transporting the bacteria through the reservoir rock would face some difficulties. Yarbrough and Coty [30] reported a field test performed by them in 1954 in Arkansas, in which Clostridium acetobutylicum was injected along with a 2% solution of beet molasses in fresh water during a 6-month period. 70 days after starting the injection, freshwater breakthrough occurred at the production well. Fermentation products such as short-chain fatty acids, CO_2, and traces of ethanol, 1-butanol, and acetone and sugars were found 80 to 90 days after the injection started. There was no increase in hydrogen content. Production of oil increased from 0.6 bbl/day to 2.1 bbl/day. Field test studies were not conducted.

The first MEOR project in the Rocky Mountains was started in 1983 [52]. An independent oil operator acquired three field service operations from Petroleum Bio-Resources Company. These were (a) a reservoir field conditioning system to avoid plugging; (b) use of a microorganism that produces gas and surfactant; and (c) use of a microorganism that produces a polysaccharide for mobility control. It was stated that production immediately doubled due to well stimulation and also increased oil recovery from 26 to 60 bbl/day, probably due to mobilization of oil by microorganisms, and water flooding was also noticed in other fields [53]. Bryant and Douglas [54] demonstrated the oil recovery efficiency of several different bacterial strains in Berea sandstone cores. They reported that additional 32% oil was recovered as compared to water flooding, and some spore-forming bacteria even showed 50–60% additional oil recovery. Berea sandstone core experiments showed that selected microbial strains could recover up to 72% of the heavy oil (API 14° and 17°) left after water flooding.

Field Tests

Kuznetsov [55] reported that bacteria were present in certain oil-gas-bearing strata in the Saratov and Buguruslan areas of the USSR in such numbers that large quantities of CO_2 were generated (depth was approximately 3,300 ft). Certainly methane was also formed. In the later works, Kuznetsov et al. [31] introduced a mixed culture of aerobic and anaerobic bacteria with acid-hydrolyzed substances from peat and soils and shut in the well for 6 months, and after that the well was opened for production [50]. The rate of oil production rose from 275 to 300 bbl/day; however, 4 months later it had fallen to 270 bbl/day. Field tests were done by Dostálek and Spurny [33] in Czechoslovakia where they injected sulfate-reducing (Desulfovibrio) and hydrocarbon-utilizing (Pseudomonas) bacteria with nutrients (molasses). During six-month experiment period, the daily average oil production increased by nearly 7%. No further work has been reported since 1958. Heningen et al. [56] reported on two field tests performed in the Netherlands, in which they used Betacoccus dextranicus in a sucrose-molasses medium of 10% total sugar content and obtained a 30% increase in cumulative oil recovery. A mixed culture of slime-forming bacteria in 50% molasses was used in the subsequent field trial. The oil-to-water production ratio changed to 1:20 compared to 1:50 before the treatment.

In Hungary, to recover naphthenic crude, Jaranyi et al. [57] utilized a mixture of anaerobic thermophilic bacteria that fermented molasses. They also tried with raw sewage as an inoculum (100L, along with 20 to 40,000 kg molasses) in their later trials (1969-70). The deepest reservoir was 8,200 ft, where the pressure was 228 atm and the temperature was 97°C. In 70% of the reservoirs tested, the introduced microbial populations showed positive results on overall oil recovery.

Karaskiewicz [58] conducted 18 field trials in Poland between 1961 and 1969. Microbial cultures were obtained from soil and water samples which were collected from the nearby areas of the oil fields and from sugar factory waters. The mixed culture includes the genera Arthrobacter, Clostridium, Mycobacterium,Peptococcus, and Pseudomonas grown in 10L bottles with formation water plus 4% molasses, incubated at 32°C. The wells ranged in depth from 1,650 to 5,000 ft. The rate of additional oil recovery ranged from 20 to 200% of the original production rate. An additional supply of nutrients was proved to be a major factor for the increased oil recovery. Lazar [59] published an extensive review of MEOR work done in Romania during the last decade, in which he discussed three major areas in MEOR including (a) isolation of the bacterial population from the formation water of the reservoir; (b) adaptation of these microorganisms in wet lab for oil release; and (c) field testing of such adapted cultures. Seven wells were treated with microbial formulations, and he concluded that the bacterial population caused an increase of oil flow up to 200% for 1 to 5 years in 2 out of 7 reservoirs (the other five were unaffected), and much information about the ecology of the reservoir is needed before initiating any MEOR activity. A list of various reported successful MEOR applications at laboratory scale and field are listed in Table 1.

Table 1: Successful laboratory and field MEOR applications [7, 13, 17–21]

Country	Biological systems used
USA	Pure or mixed cultures of Bacillus, Clostridium, Pseudomonas, and Gram-negative rods; mixed cultures of hydrocarbon degrading bacteria; mixed cultures of marine source bacteria; spore suspension of Clostridium; indigenous stratal microflora; slime-forming bacteria; ultramicrobacteria
Russia	Pure cultures of C. tyrobutiricum; bacteria mixed cultures; indigenous microflora of water injection and water formation; activated sludge bacteria; naturally occurring microbiota of industrial (food) wastes
China	Mixed enriched bacterial cultures of Bacillus, Bacteroides, Eubacterium, Fusobacterium, Pseudomonas; slime-forming bacteria: Brevibacterium viscogenes, Corynebacterium gumiform, Xanthomonas campestris
Australia	Ultramicrobacteria with surface active properties
Bulgaria	Indigenous oil-oxidizing bacteria from water injection and water formation
Canada	Pure culture of Leuconostoc mesenteroides
Former Czechoslovakia	Hydrocarbon oxidizing bacteria (predominant Pseudomonas sp.); sulfate-reducing bacteria
England	Naturally occurring anaerobic strain, high generator of acids; special starved bacteria, good producers of exopolymers
Former East Germany	Mixed cultures of thermophilic Bacillus and Clostridium from indigenous brine microflora
Hungary	Mixed sewage-sludge bacteria cultures (predominant: Clostridium, Desulfovibrio, Pseudomonas)
Norway	Nitrate-reducing bacteria naturally occurring in North Sea water
Oman	Autochthonous spore-forming bacteria from oil wells and oil contaminated soil
Poland	Mixed bacteria cultures (Arthrobacter, Clostridium, Mycobacterium, Peptococcus, Pseudomonas)
Romania	Adapted mixed enrichment cultures (predominant: Bacillus, Clostridium, Pseudomonas, and other Gram-negative rods)
Saudi Arabia	Adequate bacterial inoculum according to requirements of each technology
The Netherlands	Slime-forming bacteria (Betacoccus dextranicus)
Trinidad-Tobago	Facultative anaerobic bacteria high producers of gases
Venezuela	Adapted mixed enrichment cultures

HEAVY OIL

Heavy crude oil or extra heavy crude oil is a type of crude oil which does not flow easily. It is referred to as "heavy" because of its density or specific gravity, which is higher than that of light crude oil. Heavy crude oil has been defined as any liquid petroleum with API gravity less than 20°, which means its specific gravity is greater than 0.933. This type of oil forms due to the exposure of crude oil to bacteria [60]. Production, transportation, and refining of heavy crude oil are much difficult compared to light crude oil. The largest reserves of heavy oil in the world are located in the north of the Orinoco River in Venezuela (Energy Information Administration, 2001) the same amount as the conventional oil reserves of Saudi Arabia, but 30 or more countries are known to have such heavy crude oil reserves. Heavy crude oil is closely related to oil sands; the main difference is that oil sands generally do not flow at all. Canada has large reserves of oil sands, located north and northeast of Edmonton, Alberta. Physical properties that distinguish heavy crudes from lighter ones include higher viscosity and specific gravity, as well as heavier molecular composition. Extra heavy oil from the Orinoco region has a viscosity of over 10,000 centipoise and 10° API gravity. A diluent is added at regular distances in pipeline carrying heavy oil to increase the flow rate [61].

Field Tests

Heavy crude oil plays a major role in the economics of petroleum development. The heavy oil resources in the world are more than twice those of conventional light crude oil. In October 2009, the USGS updated the Orinoco tar sands (Venezuela) recoverable value to 513 billion barrels ($8.16 \times 1010 \, m^3$) (USGS. 11 January 2010), making this area the world's first recoverable oil deposit, ahead of Saudi Arabia and Canada [61]. The price of heavy crude oil slashes as compared to light oil due to increased refining costs and high sulphur content. The high viscosity and density also make production more difficult. On the other hand, large quantities of heavy crudes have been discovered in the Americas including Canada, Venezuela, and California. Another reason can be the relatively shallow depth of heavy oil fields (often less than 3000 feet) which contributes to lower production costs [62].

Special techniques are being developed for exploration and production of heavy oil.

Chemical Properties

Heavy oil contains asphaltenes and resins. It is "heavy" (dense and viscous) due to the high ratio of aromatics and naphthenes to paraffins (linear alkanes) and high amounts of NSOs (nitrogen, sulfur, oxygen, and heavy metals). The carbon chain in heavy oil has over 60 carbon atoms which results in a high boiling point and molecular weight. For example, the viscosity of Venezuela's Orinoco extra-heavy crude oil lies in the range of 1000–5000 cP (1–5 Pa·s), while Canadian extra-heavy crude has a viscosity in the range of 5000–10,000 cP (5–10 Pa·s), about the same as molasses, and higher (up to 100,000 cP or 100 Pa·s for the most viscous commercially exploitable deposits) [62]. A definition from the Chevron Phillips Chemical Company is as follows.

The "heaviness" of heavy oil is primarily the result of a relatively high proportion of a mixed bag of complex, high molecular weight, nonparaffinic compounds, and a low proportion of volatile, low molecular weight compounds. Heavy oils typically contain very little paraffin and may or may not contain high levels of asphaltenes.

DEVELOPMENT OF HEAVY OIL RESERVES IN OMAN

The first oil discovery in the Sultanate of Oman was accomplished in 1956, when City Services Company drilled Marmul-1 well. But the discovery was not considered as a commercial discovery because the oil found was heavy compared to oil discoveries in the Middle East at that time. In 1962, Petroleum Development of Oman (PDO) exploration activities ended up in achieving commercial discovery of oil in Yibal field, followed by discoveries in Natih and Fahud fields in 1963 and 1964, respectively. These discoveries marked the birth of Oman as an oil producing country. The result of these successes in discovering and production of oil inspired the Government to sign two new agreements to explore oil and gas in 1973 and another two in 1975 with other international oil companies. By 2009, the number of active oil fields reached 135. Over these days, Oman has been continuously applying efforts to improve the recovery of its oil reserves and has adopted EOR techniques on a large scale. These initiatives helped the Sultanate to increase their oil production capability to nearly 1 million barrels per day (bpd) from 714300 bpd averaged in 2008 [63]. This has also changed the outlook for its oil industry which is now estimated to have at least 40 years of life ahead of it [64]. Al-Ghubar South's discovery in 2009 was the most auspicious discovery for Oman. According to the Ministry of Oil and Gas, this discovery could add as much as 1 billion barrels to reserves. Two other convincing discoveries, including that in Malaan West and Taliah in the Lekhwair cluster in northwest Oman, were made which will stretch the baseline production in the future [65].

First Trials

The first export of Omani oil took place on July 27,1967. In the beginning, oil production increased steadily to 341000 barrels per day in 1975 and in 1984; the average daily production reached around 400000 barrels per day. Petroleum Development Oman (PDO)—the largest oilfield operator in Oman—started a series of EOR trials in 1986 due to low recovery of oil because of the complex geology of the reservoirs. The trials proved successful and Oman slowly started implementing EOR thereby boosting the production to a current level of nearly 900000 bpd. EOR projects result in 5–15 percent increment in reserves and PDO expects its EOR projects to contribute around 35 percent of its total production by 2020. So Oman is considered as a country which is pushing the limits of EOR technology [66].

Miscible Gas Injection

Miscible gas injection involves pumping gas to oil wells. These gases that are being used for this purpose are often toxic which will dissolve in the oil and eventually lead to higher flow rates. This technique is currently at its operations in the Harweel oil field cluster [65].

Steam Injection

Qarn Alam is the world's first full field EOR project and also the largest of its kind in the world. Thermally assisted gas oil gravity drainage (TAGOGD), a sophisticated method, is employed due to the characteristics of the fractured carbonate reservoir, as the oil is highly viscous and a very low percentage of recovery is feasible by conventional oil extraction method.

Polymer Injection

Marmul field is located in south Oman. It is characterized by heavy viscous crude that is difficult to extract by traditional recovery methods. The reservoir has a viscosity of around 90 cP. The reservoir's sweep efficiency was modified by viscosifying the water with the addition of polyacrylamide polymers and then injected in the reservoir through polymer injection wells. The polymer flooding at Marmul field will increase a further recovery by 8000 bpd. By this technique, 10–15% increase in recovery levels from the Marmul reservoirs is predicted.

EOR Projects in Oman Oil Fields

Miscible gas injection has been applied in Harweel oil field which resulted in an additional production of 40,000 bbl/day. Thermal EOR methods are being deployed at Mukhaizna, Marmul, Amal-East, Amal-West, and Qarn Alam fields. Mukhaizna has already increased production to 50,000 bbl/day, and the other fields, Amal-East and Amal-West, are expected to raise the production to 23,000 bbl/day by 2018. Furthermore, the steam injection at Qarn Alam is supposed to enhance the production by 40,000 bbl/day by 2015. This is achieved by a novel process in which steam drains oil to lower producer wells. At projects such as Marmul, with its heavy oil reserves, injecting polymer fluid has seen to be more effective.

Other EOR projects include Karim cluster, a cluster of 18 oil fields flowing to the Nimr production facility, in which PDO is aiming to boost up the production. In Harweel cluster, PDO estimates approximately 40 percentage increase in the next five years. Also with Rima clusters, using EOR techniques, much gain is expected (US Energy Information Administration, 2012).

ROLE OF MICROBES IN BIODEGRADATION OF HEAVY CRUDE OIL

Degradation of oil is one of the most important parts of the MEOR by which the oil's viscosity and freezing point are reduced which in turn will increase the oil's flow in situ. Heavy oils are rich in gum and asphaltene, having characteristics such as freezing point, low flow ability, difficult oil recovery, and high recovery cost [67].

Microbe can improve the physical characteristics of heavy oil in two ways: (1) by degrading heavy oil fractions, thereby decreasing the average molecular weight of heavy oil; and (2) the byproducts of microbial metabolism, such as biological surface active substance, acid, and gas, which can reduce the viscosity of oil considerably. Gum and asphaltene present in heavy oil have high molecular weight and polarity; meanwhile they are one of the main factors making the oil recovery difficult [68]. Usually, microbe hardly degrades them [69]. Zhang et al. [70] separated a variety of microbes from environments rich in petroleum and done a series of experiments using mixed microbial consortia, which can effectively degrade heavy oil, even gum and asphaltene; these microbes act by lowering the viscosity and freezing point of heavy oil and thereby improving the physical and chemical characters of heavy oils.

In some cases, using microbial consortia with different properties (ability to degrade heavy oil fractions and biosurfactant production) thereby applying different mechanisms might have a desired effect for enhanced oil recovery [71]. There are a lot of microbes having the ability to degrade hydrocarbons by using them as carbon sources [72]. Interesting results for the microbial n-alkane degradation have been reported during the past decades [73–76]. Extensive studies have been made on strains of Gordonia amicalis which have shown to be a potent degrader of large n-alkanes under aerobic and anaerobic conditions [77]; manyPseudomonas species have the ability to degrade lighter hydrocarbons with carbon chain length C_{12}–C_{32}, and heavier hydrocarbons with carbon chain length of C_{36}–C_{40} [78, 79]; and a thermophilic Bacillus strain that degrades only long-chain (C_{15}–C_{36}) hydrocarbons but not short-chain (C_{8}–C_{14}) n-alkanes [80] has also been reported. The ability of biosurfactant-producing indigenous Bacillus strains to degrade the higher fractions of crude oil and aid in the enhancement of its flow characteristics has also been studied for a petroleum reservoir in the Daqing Oilfield [81]. The MEOR team in the Sultan Qaboos University, Oman, found that a consortia of Bacillus strains form oil contaminated soil degraded heavy chain oil (C_{50}–C_{70}) to (C_{11}–C_{20}). Many microorganisms contain genes coding for the enzymes responsible for degrading petroleum hydrocarbons. Some microorganisms degrade alkanes (normal, branched, and cyclic paraffins), others aromatics, and others both paraffinic and aromatic hydrocarbons [82–84]. The most readily degraded alkanes are considered to be in in the range of C_{10} to C_{26}, but low-molecular-weight aromatics, such as benzene, toluene, and xylene, which are considered as the toxic compounds found in petroleum, are also readily biodegraded by many marine microorganisms. As the complexity of the structures (those with branches and/or condensed ring structures) increases, it will be more resistant for biodegradation, which means only fewer microorganisms can degrade those structures and the biodegradation rates would be much lower than the rates for the simpler hydrocarbon structures found in petroleum. The higher the number of methyl-branched components or condensed aromatic rings, the slower the rates of biodegradation and the greater the probability of accumulating partially oxidized intermediary metabolites.

Petroleum contains numerous compounds of varying structural complexities. The residual mixture formed after petroleum biodegradation may resist further biodegradation. Crude oils are never completely degraded and always result in some complex residue which appears as a black tar containing a high proportion of asphaltic compounds. The toxicity and bioavailability of the residual mixture are very low as long as it does not coat and suffocate an area, thus becoming an inert environmental contaminant with no toxic effects on environment [60].

About 10% of the total bacterial population in hydrocarbon-contaminated marine environments is hydrocarbon-degrading bacterial populations [82]. The major metabolic pathways for hydrocarbon biodegradation have been elucidated [85]. The initial steps in the biodegradation of hydrocarbons by bacteria are the oxidation of the oil by oxygenases. Alkanes are subsequently converted to carboxylic acids that are further biodegraded via β-oxidation (the central metabolic pathway for the utilization of fatty acids from lipids, which results in the formation of acetate, enters into the tricarboxylic acid cycle). Generally aromatic hydrocarbon rings are hydroxylated to form diols, which are then eventually cleaved to form catechols which are subsequently degraded to intermediates of the tricarboxylic acid cycle. Interestingly, the intermediates resulting from bacterial action are with differing stereochemistry usually cis-diols, which are biologically inactive. With bacteria being the dominant hydrocarbon degraders in the marine environment, the products of aromatic hydrocarbons biodegradation will detoxify them and do not produce potential carcinogens. The complete biodegradation (mineralization) of hydrocarbons produces environmentally safe end products such as carbon dioxide and water, as well as cell biomass (largely protein) which will eventually enter into the food web.

MICROBIAL CANDIDATES INVOLVED IN CRUDE OIL DEGRADATION

Thermophilic Spore-Forming Bacteria Involved in Biodegradation of Heavy Crude Oil for MEOR

Many varieties of microbes are identified and isolated from different petroleum reservoirs which comprise several ecological niches, including sulfate reducers [86–88], sulfur reducers [86], methanogen [88], fermentative bacteria [89, 90], manganese and iron reducers [91], and dibenzothiophene-degrading bacteria [92]. Although many bacteria are isolated from many reservoirs, those which can be applied to MEOR are fewer.

Many researchers have been engaged in studying thermophiles. It is reported that 140 species of 70 genera of thermophiles have been discovered from high temperature environments with wide applications [93]. In the Shengli oil field of East China, where extreme physical conditions exist with temperature ranging 60–90°C and depth of 1000–2000 m, most of the reservoirs are under EOR. This harsh environment seems to be unsuitable for microbial growth. But some thermophiles have been isolated which helps in EOR [86].

There are many kinds of Bacillus, which are distributed widely, but those which have application on crude oil recovery are very few [94, 95]. B. subtilis and B. licheniformis strains have been repeatedly isolated from many oil reservoirs as well as oil contaminated samples, thus confirming the adaptability of these species [17–19, 95–101]. The properties of B. subtilis have been reported in much literature [17–19, 96, 102–104], but the isolation and its action on crude oil have been scarcely reported [95].

It is recognized that the thermophiles possess enzymes which are more resistant to physical and chemical denaturation. Their faster growth rates also serve as another major advantage. Relative studies suggest that thermophilic hydrocarbon degraders of Bacillus, Thermus, Thermococcus, and Thermotoga species occurring in natural high-temperature or sulfur-rich environments are of special significance [105]. Wang et al. [106] isolated functional bacteria from high temperature petroleum reservoirs. Three thermophilic hydrocarbon-degrading bacteria, which belonged to Bacillus sp., Geobacillus sp., and Petrobacter sp., could tolerate 55°C in obligate anaerobic condition. These strains could utilize crude oil as carbon source with the degradation rate of 56.5%, 70.01%, and 31.78%, respectively, along with the viscosity reduction rate of 40%, 54.55%, and 29.09%, meanwhile the solidify points of crude oil were reduced by 3.7, 5.2, and 3.1°C.

Hao et al. [107] isolated a hydrocarbon-degrading bacterium, strain SB-1, from oil-contaminated soil samples collected at the Shengli oil field in east China. Based on 16S rDNA sequence, the strain was identified as B. subtilis. The bacteria degraded 39.33% of crude oil, 57.01% of the saturated fractions, 25.63% of the resins, and 12.15% of the aromatic fractions within 12 days. In addition, more than 50% of the alkanes were removed by the strain; the highest degradation rate was shown as 81.03% for C_{36}–C_{40}, and the lowest degradation rate being 51.47% for C_{31}–C_{35}. The results of this study concluded that B. subtilis SB-1 is a potent strain in degrading oil pollutants in soil.

Sanchez et al. [108] isolated thermophilic bacteria enriched from the formation waters of a Venezuelan oil field. The reservoir, located at Maracaibo Lake, has a temperature of 60–80°C and a pressure of 1,200–1,500 psi. The main fermentative byproducts were alcohols, short chain fatty acids, and gases when grown in media with industrial wastes as carbon source.

A strain of B. stearothermophilus (Geobacillus) was isolated from oil-contaminated Kuwaiti desert capable of growing on C_{15}–C_{17} [109], and two strains of G. jurassicus were isolated from a high temperature petroleum reservoir capable of growing on C_6–C_{16} [110]. B. thermoleovorans strain isolated from deep subterranean petroleum reservoirs was shown to degrade n-alkane up to C_{23} at 70°C [111]. Thermophilic, glucose-fermenting, strictly anaerobic, rod-shaped bacterium, Thermotoga hypogea sp. strain SEBR 6459T (T = type strain), was isolated from an African oil-producing well [112] and T. elfii strain SEBR 6459 by Ravot et. al. [113]. Al-Bahry et al. [18–21, 96] reported 33 genera and 58 species identified from Omani oil wells. All of the identified microbial genera were first reported in Oman, with Caminicella sporogenes for the first time reported from oil fields. Most of the identified microorganisms were found to be anaerobic, thermophilic, and halophilic and produced biogases, biosolvents, and biosurfactants as by-products, which may be potentially applicable in MEOR.

Various bioremediation and biodegradation agents are commercially available consisting of microbial cultures or microbial enzymes or both. The US Environmental Protection Agency National Contingency Plan released a product schedule report on August 2013 [114]. Also various laboratory screening reports are available for these commercial products [115].

CONCLUSIONS

Given the scarcity of the literature on thermophilic spore-forming bacteria involved in MEOR for crude oil biodegradation, there is a clear need for further laboratory research. While significant progress has been made, we still need to rigorously examine this mechanism of MEOR.

REFERENCES

1.   K. Aleklett, M. Höök, K. Jakobsson, M. Lardelli, S. Snowden, and B. Söderbergh, "The peak of the oil age—analyzing the world oil production reference scenario in world energy outlook 2008," Energy Policy, vol. 38, no. 3, pp. 1398–1414, 2010.

2.   W. Graus, M. Roglieri, P. Jaworski, L. Alberio, and E. Worrell, "The promise of carbon capture and storage: evaluating the capture-readiness of new EU fossil fuel power plants," Climate Policy, vol. 11, no. 1, pp. 789–812, 2011.

3.   R. Cossé, Basics of Reservoir Engineering, Pure and Applied Geophysics, Éditions Technip, 1993.

4.   C. Hall, P. Tharakan, J. Hallock, C. Cleveland, and M. Jefferson, "Hydrocarbons and the evolution of human culture," Nature, vol. 426, no. 6964, pp. 318–322, 2003.

5.   R. S. Bryant, A. K. Stepp, K. M. Bertus, T. E. Burchfield, and M. Dennis, "Microbial-enhanced waterflooding field pilots," Developments in Petroleum Science, vol. 39, pp. 289–306, 1993. · ·

6.   L. W. Lake, Enhanced Oil Recovery, Prentice Hall, Englewood Cliffs, NJ, USA, 1989.

7.   I. Lazar, I. G. Petrisor, and T. F. Yen, "Microbial enhanced oil recovery (MEOR)," Petroleum Science and Technology, vol. 25, no. 11, pp. 1353–1366, 2007. · ·

8.   R. Sen, "Biotechnology in petroleum recovery: the microbial EOR," Progress in Energy and Combustion Science, vol. 34, no. 6, pp. 714–724, 2008. · ·

9.   D. O. Hitzman, "Microbial enhanced oil recovery—the time is now," Developments in Petroleum Science, vol. 31, pp. 11–20, 1991. · ·

10.   B. Bubela, "A comparison of strategies for enhanced oil recovery using in situ and ex situ produced biosurfactants," Surfactant Science Series, vol. 25, pp. 143–161, 1987.

11.   G. S. Derek, "Microbiological methods for the enhancement of oil recovery," Biotechnology and Genetic Engineering Reviews, vol. 1, no. 1, pp. 187–222, 1984. ·

12.   S. Rebeka, "Potential uses of microorganisms in petroleum recovery technology," in Proceedings of the Oklahoma Academy of Science, 1987.

13.   R. Al-Hattali, H. Al-Sulaimani, Y. Al-Wahaibi et al., "Microbial biomass for improving sweep efficiency in fractured carbonate reservoir using date molasses as renewable feed substrate," inProceedings of the SPE Annual Technical Conference and Exhibition, San Antonio, Tex, USA, 2012.

14. H. Suthar, K. Hingurao, A. Desai, and A. Nerurkar, "Evaluation of bioemulsifier mediated microbial enhanced oil recovery using sand pack column," Journal of Microbiological Methods, vol. 75, no. 2, pp. 225–230, 2008. ··

15. I. M. Banat, A. Franzetti, I. Gandolfi et al., "Microbial biosurfactants production, applications and future potential," Applied Microbiology and Biotechnology, vol. 87, no. 2, pp. 427–444, 2010. ··

16. A. K. Sarkar, J. C. Goursaud, M. M. Sharma, and G. Georgiou, "Critical evaluation of MEOR processes," In Situ, vol. 13, no. 4, pp. 207–238, 1989.

17. S. N. Al-Bahry, Y. M. Al-Wahaibi, A. E. Elshafie et al., "Biosurfactant production by Bacillus subtilisB20 using date molasses and its possible application in enhanced oil recovery," International Biodeterioration and Biodegradation, vol. 81, pp. 141–146, 2013. ··

18. H. Al-Sulaimani, Y. Al-Wahaibi, S. N. Al-Bahry et al., "Experimental investigation of biosurfactants produced by Bacillus species and their potential for MEOR in Omani oil field," in Proceedings of the SPE EOR Conference at Oil and Gas West Asia 2010 (OGWA ‹10), pp. 378–386, Muscat, Oman, April 2010.

19. H. Al-Sulaimani, Y. Al-Wahaibi, S. Al-Bahry et al., "Optimization and partial characterization of biosurfactants produced by Bacillus species and their potential for ex-situ enhanced oil recovery," SPE Journal, vol. 16, no. 3, pp. 672–682, 2011.

20. H. Al-Sulaimani, Y. Al-Wahaibi, S. N. Al-Bahry et al., "Residual-oil recovery through injection of biosurfactant, chemical surfactant, and mixtures of both under reservoir temperatures: induced-wettability and interfacial-tension effects," SPE Reservoir Evaluation and Engineering, vol. 15, no. 2, pp. 210–217, 2012.

21. Y. Al-Wahaibi, H. Al-Hadrami, S. Al-Bahry, A. Elshafie, A. Al-Bemani, and S. Joshi, "Residual oil recovery via injection of biosurfactant and chemical surfactant following hot water injection in Middle East heavy oil field," in Proceeding of the SPE Heavy Oil Conference, Alberta, Canada, June 2013.

22. K. Fujiwara, Y. Sugai, N. Yazawa, K. Ohno, C. X. Hong, and H. Enomoto, "Biotechnological approach for development of microbial enhanced oil recovery technique," Studies in Surface Science and Catalysis, vol. 151, pp. 405–445, 2004.

23. H. Al-Sulaimani, S. Joshi, Y. Al-Wahaibi, S. N. Al-Bahry, A. Elshafie, and A. Al-Bemani, "Microbial biotechnology for enhancing oil recovery: current developments and future prospects," Biotechnology, Bioinformatics and Bioengineering Journal, vol. 1, no. 2, pp. 147–158, 2011.

24. A. R. Awan, R. Teigland, and J. Kleppe, "A survey of North Sea enhanced-oil-recovery projects initiated during the years 1975 to 2005," SPE Reservoir Evaluation and Engineering, vol. 11, no. 3, pp. 497–512, 2008.

25. J. W. Beckman, "The action of bacteria on mineral oil," Industrial and Engineering Chemistry, News Edition, vol. 4, pp. 23–26, 1926.

26. C. E. Zobell, "Bacterial release of oil from oil-bearing materials," World Oil, vol. 126, pp. 36–47, 1947.

27. D. M. Updegraff and G. B. Wren, "The release of oil from petroleum-bearing materials by sulfate-reducing bacteria," Applied Microbiology, vol. 2, no. 6, pp. 309–322, 1954.

28. J. B. Davis and D. M. Updegraff, "Microbiology in the petroleum industry," Bacteriological Reviews, vol. 18, no. 4, pp. 215–238, 1954.

29. D. M. Updegraff, "Recovery of petroleum oil," US Patent No. 2.807.570, 1957.

30. H. F. Yarbrough and V. F. Coty, "Microbial enhanced oil recovery from the upper crustaceous nacatoch formation," in Proceedings of the International Conference on Microbial Enhancement of Oil Recovery, 1983.

31. S. I. Kuznetsov, M. V. Ivanov, and N. N. Lyalikowa, Introduction to Geological Microbiology, McGraw-Hill, New York, NY, USA, 1963.

32. M. V. Ivanov, S. S. Belyaev, M. A. Zyakun, A. V. Bondar, and S. K. Laurinavichus, "Microbiological formation of methane in the oil field development," Moscova, vol. 11, 1983.

33. M. Dostálek and M. Spurny, "Bacterial release of oil. A preliminary trial in an oil deposit," Folia Biologica, vol. 4, pp. 166–172, 1958.

34. M. Dienes and I. Yaranyi, "Increase of oil recovery by introducing anaerobic bacteria into the formation Demjen field," Hungary Koolaj as Fodgas, vol. 106, no. 7, pp. 205–208, 1973.

35. I. Karaskiewicz, "The application of microbiological method for secondary oil recovery from the Carpathian crude oil reservoir," Widawnistwo "SLASK", pp. 1–67, 1974.

36. D. O. Hitzman, "Review of microbial enhanced oil recovery field tests," in Proceedings of the Applications of Microorganisms to Petroleum Technology, U.S. Department of Energy, 1988.

37. I. Lazar, "MEOR field trials carried out over the world during the past 35 years," in Microbial Enhancement of Oil Recovery— Recent Advances, E. C. Donaldson, Ed., 1991.

38. I. Lazar, "International MEOR applications for marginal wells," Pakistan Journal of Hydrocarbon Research, vol. 10, pp. 11–30, 1998.

39. M. V. Ivanov, S. S. Belyaev, I. A. Borzenkov, I. F. Glumov, and P. B. Ibatulin, "Additional oil production during field trials in Russia," in Microbial Enhancement of Oil Recovery—Recent Advances, E. Premuzic and A. Woodhead, Eds., 1993.

40. J. E. Zajic, D. G. Cooper, T. R. Jack, and N. Kosaric, Microbial Enhanced Oil Recovery, Penn Well Books, Tulsa, Okla, USA, 1983.

41. T. F. Yen, State of the Art Review on Microbial Enhanced Oil Recovery, NSF OIR-8405134, Los Angeles, Calif, USA, 1986.

42. E. C. Donaldson, G. V. Chilingarian, and T. F. Yen, Microbial Enhanced Oil Recovery, Elsevier, New York, NY, USA, 1989.

43. E. A. Grula, H. H. Russell, D. Bryant, M. Kanaga, and M. Hart, "Isolation and screening of Clostridia for possible use in microbially enhanced oil recovery," in Proceedings of the Microbial Enhanced Oil Recovery, Afton, Okla, USA, 1982.

44. C. E. Zobell, "Bacteriological process for treatment of fluid-bearing earth formations," US Patent No. 2, 413, 278, 1946.

45. S. Joshi, C. Bharucha, S. Jha, S. Yadav, A. Nerurkar, and A. J. Desai, "Biosurfactant production using molasses and whey under thermophilic conditions," Bioresource Technology, vol. 99, no. 1, pp. 195–199, 2008. · ·

46. D. C. Bond, Bacteriological Method of Oil Recovery, Pure Oil Company, Rock Hill, SC, USA, 1961.

47. D. O. Hitzman, "Microbiological secondary recovery of oil," U.S. Patent 3, 032, 472, 1962.

48.   R. M. Knapp, M. J. McInerney, D. E. Menzie, and G. E. Jenneman, "The use of microorganisms in enhanced oil recovery," in First Annual Report to the Department of Energy, 1983.

49.   A. C. Johnson, "Microbial oil release technique for enhanced oil recovery," in Proceedings of the Conference on Microbiological Processes Useful in Enhanced Oil Recovery, San Diego, Calif, USA, 1979.

50.   D. O. Hitzman, "Petroleum microbiology and the history of its role in enhanced oil recovery," inProceedings of the International Conference on Microbial Enhancement of Oil Recovery, U. S. Department of Energy, Bartlesville, Okla, USA, 1982.

51.   T. R. Jack, B. G. Thompson, and E. D. Blasio, "The potential for use of microbes in the production of heavy oil," in Proceedings of the International Conference on Microbial Enhanced Oil Recovery, Afton, Okla, USA, 1982.

52.   R. Rountree, "Rocky mountain oil history," Western Oil Reporter, vol. 4, p. 77, 1984.

53.   J. E. Zajic, Proceedings of the 1st International MEOR Workshop, U.S. Department of Energy Report No. DOE/BC/10852-1, 1986.

54.   R. S. Bryant and J. Douglas, "Evaluation of Microbial Systems in Porous Media for EOR," SPE Reservoir Engineering, vol. 3, no. 2, pp. 489–495, 1988.

55.   S. I. Kuznetsov, "Possibilities of production of methane in oil fields of Saratov and Buguruslan,"Mikrobiologiia, vol. 19, no. 3, pp. 193–202, 1950.

56.   J. V. Heningen, A. J. DeHann, and J. D. Jansen, "Process for the recovery of petroleum from rocks," Netherlands Patent 80, 580, 1958.

57.   I. Jaranyi, L. Kiss, C. Salanczy, and J. Szolnoki, "Alteration of some characteristics of oil-wells through the effects of microbial treatment," in Proceedings of the 3rd International Science Conference on Geochemistry, 1963.

58.   J. Karaskiewicz, "Studies on increasing petroleum oil recovery from Carpathian deposits using bacteria," Nafta (Petroleum), vol. 21, pp. 144–149, 1975.

59.   I. Lazar, "Microbially enhanced oil recovery in Romania," in Proceedings of the International Conference on Microbial Enhanced Oil Recovery, Afton, Okla, USA, 1982.

60. M. B. Dusseault, "Comparing Venezuelan and Canadian heavy oil and tar sands," in Proceedings of the Petroleum Society›s Canadian International Petroleum Conference, Calgary, Canada.

61. H. A. Rodriguez, P. Vaca, O. Gonzalez, and M. C. de Mirabal, "Integrated study of a heavy oil reservoir in the Orinoco Belt: a field case simulation," in Proceedings of the SPE Reservoir Simulation Symposium, pp. 309–310, June 1997.

62. S. Chopra and L. Lines, "Introduction to this special section: heavy oil," The Leading Edge, vol. 27, no. 9, pp. 1104–1106, 2008. ·

63. http://www.engineerlive.com/content/23933.

64. http://www.mog.gov.om/english/AboutUs/TheHistoryofOilGas/tabid/117/Default.aspx.

65. http://www.eia.gov/countries/cab.cfm?fips=MU.

66. http://www.reportlinker.com/p0155667/oman-oil-and-Gas-Report-Q4.html.

67. G. L. Lei, "Research and application of microbial enhanced oil recovery," Acta Petrolei Sinica, vol. 22, no. 2, pp. 56–61, 2001.

68. Y. S. Peng, H. S. Ji, and C. X. Liang, Field Research of Microbial Enhanced Oil Recovery, Petroleum Industry Press, Beijing, China, 1997.

69. T. S. Zhang, G. Z. Lan, L. Deng, and X. G. Deng, "Experiments on heavy oil degradation and enhancing oil recovery by microbial treatments," Acta Petrolei Sinica, vol. 22, no. 1, pp. 54–57, 2001.

70. T. S. Zhang, X. Chen, G. Z. Lan, and Z. Jiang, "Microbial degradation influences on heavy oil characters and MEOR test," in Proceedings of the 18th World Petroleum Congress, Johannesburg, South Africa, September 2005.

71. L. Jinfeng, M. Lijun, M. Bozhong, L. Rulin, N. Fangtian, and Z. Jiaxi, "The field pilot of microbial enhanced oil recovery in a high temperature petroleum reservoir," Journal of Petroleum Science and Engineering, vol. 48, no. 3-4, pp. 265–271, 2005. · ·

72. A. Wentzel, T. E. Ellingsen, H. K. Kotlar, S. B. Zotchev, and M. Throne-Holst, "Bacterial metabolism of long-chain n-alkanes," Applied Microbiology and Biotechnology, vol. 76, no. 6, pp. 1209–1221, 2007. · ·

73. V. G. Grishchenkov, R. T. Townsend, T. J. McDonald, R. L. Autenrieth, J. S. Bonner, and A. M. Boronin, "Degradation of petroleum hydrocarbons by facultative anaerobic bacteria under aerobic and anaerobic conditions," Process Biochemistry, vol. 35, no. 9, pp. 889–896, 2000. · ·

74. J. S. Sabirova, M. Ferrer, D. Regenhardt, K. N. Timmis, and P. N. Golyshin, "Proteomic insights into metabolic adaptations in Alcanivorax borkumensis induced by alkane utilization," Journal of Bacteriology, vol. 188, no. 11, pp. 3763–3773, 2006. · ·

75. A. Etoumi, I. El Musrati, B. El Gammoudi, and M. El Behlil, "The reduction of wax precipitation in waxy crude oils by Pseudomonas species," Journal of Industrial Microbiology and Biotechnology, vol. 35, no. 11, pp. 1241–1245, 2008. · ·

76. M. Binazadeh, I. A. Karimi, and Z. Li, "Fast biodegradation of long chain n-alkanes and crude oil at high concentrations with Rhodococcus sp. Moj-3449," Enzyme and Microbial Technology, vol. 45, no. 3, pp. 195–202, 2009. · ·

77. D. H. Hao, J. Q. Lin, X. Song, J. Lin, Y. J. Su, and Y. B. Qu, "Isolation, identification, and performance studies of a novel paraffin-degrading bacterium of Gordonia amicalis LH3," Biotechnology and Bioprocess Engineering, vol. 13, no. 1, pp. 61–68, 2008. · ·

78. I. M. Banat, "Biosurfactants production and possible uses in microbial enhanced oil recovery and oil pollution remediation: a review," Bioresource Technology, vol. 51, no. 1, pp. 1–12, 1995. · ·

79. M. Hasanuzzaman, A. Ueno, H. Ito et al., "Degradation of long-chain n-alkanes (C36 and C40) byPseudomonas aeruginosa strain WatG," International Biodeterioration and Biodegradation, vol. 59, no. 1, pp. 40–43, 2007. · ·

80. L. Wang, Y. Tang, S. Wang et al., "Isolation and characterization of a novel thermophilic Bacillusstrain degrading long-chain n-alkanes," Extremophiles, vol. 10, no. 4, pp. 347–356, 2006. · ·

81. Y. H. She, F. Zhang, J. J. Xia et al., "Investigation of biosurfactant-producing indigenous microorganisms that enhance residue oil recovery in an oil reservoir after polymer flooding," Applied Biochemistry and Biotechnology, vol. 163, no. 2, pp. 223–234, 2011. · ·

82. R. M. Atlas, "Microbial degradation of petroleum hydrocarbons: an environmental perspective,"Microbiological Reviews, vol. 45, no. 1, pp. 180–209, 1981.

83. J. G. Leahy and R. R. Colwell, "Microbial degradation of hydrocarbons in the environment,"Microbiological Reviews, vol. 54, no. 3, pp. 305–315, 1990.

84. R. M. Atlas and R. Bartha, "Hydrocarbon biodegradation and oil spill bioremediation," Advances in Microbial Ecology, vol. 12, pp. 287–338, 1992. ·

85. R. M. Atlas, Petroleum Microbiology, Macmillan, New York, NY, USA, 1984.

86. K. O. Stetter, R. Huber, E. Blöchl et al., "Hyperthermophilic archaea are thriving in deep North Sea and Alaskan oil reservoirs," Nature, vol. 365, no. 6448, pp. 743–745, 1993. · ·

87. C. Tardy-Jacquenod, P. Caumette, R. Matheron, C. Lanau, O. Arnauld, and M. Magot, "Characterization of sulfate-reducing bacteria isolated from oil-field waters," Canadian Journal of Microbiology, vol. 42, no. 3, pp. 259–266, 1996.

88. T. K. Ng, P. J. Weimer, and L. J. Gawel, "Possible nonanthropogenic origin of two methanogenic isolates from oil producing wells in the San Miguelito field, Ventura County, California,"Geomicrobiology Journal, vol. 7, no. 3, pp. 185–192, 1989. ·

89. M. E. Davey, W. A. Wood, R. Key, K. Nakamura, and D. A. Stahl, "Isolation of three species of Geotoga and Petrotoga: two new genera, representing a new lineage in the bacterial line of descent distantly related to the 'Thermotogales'," Systematic and Applied Microbiology, vol. 16, no. 2, pp. 191–200, 1993.

90. G. S. Grassia, K. M. McLean, P. Glénat, J. Bauld, and A. J. Sheehy, "A systematic survey for thermophilic fermentative bacteria and archaea in high temperature petroleum reservoirs," FEMS Microbiology Ecology, vol. 21, no. 1, pp. 47–58, 1996. · ·

91. A. C. Greene, B. K. C. Patel, and A. J. Sheehy, "Deferribacter thermophilus gen. nov., sp. nov., a novel thermophilic manganese- and iron-reducing bacterium isolated from a petroleum reservoir,"International Journal of Systematic Bacteriology, vol. 47, no. 2, pp. 505–509, 1997.

92. A. Bahrami, S. A. Shojaosadati, and G. Mohebali, "Biodegradation of dibenzothiophene by thermophilic bacteria," Biotechnology Letters, vol. 23, no. 11, pp. 899–901, 2001. · ·

93. Z. He, Q. Peng, and J. Chen, Biology of Thermophiles, Scientific Press, Beijing, China, 2000.

94. J. Li, B. Lian, J. Hao, J. Zhao, and L. Zhu, "Non-parallelism between the effect of microbial flocculants on sewerage disposal and the flocculation rate," Chinese Journal of Geochemistry, vol. 25, no. 2, pp. 139–142, 2006. · ·

95. G. E. Jenneman, M. J. McInerney, R. M. Knapp et al., "A halotolerant, biosurfactant producingBacillus species potentially useful for enhanced oil recovery," Developments in Industrial Microbiology, vol. 24, pp. 485–492, 1983.

96. S. N. Al-Bahry, A. Elshafie, Y. Al-Wahaibi et al., "Microbial consortia in Oman oil fields: a possible use in enhanced oil recovery," Journal of Microbiologyand Biotechnology, vol. 23, no. 1, pp. 106–117, 2013.

97. M. M. Yakimov, K. N. Timmis, V. Wray, and H. L. Fredrickson, "Characterization of a new lipopeptide surfactant produced by thermotolerant and halotolerant subsurface Bacillus licheniformisBAS50," Applied and Environmental Microbiology, vol. 61, no. 5, pp. 1706–1713, 1995.

98. S. M. M. Dastgheib, M. A. Amoozegar, E. Elahi, S. Asad, and I. M. Banat, "Bioemulsifier production by a halothermophilic Bacillus strain with potential applications in microbially enhanced oil recovery," Biotechnology Letters, vol. 30, no. 2, pp. 263–270, 2008. · ·

99. H. Ghojavand, F. Vahabzadeh, M. Mehranian et al., "Isolation of thermotolerant, halotolerant, facultative biosurfactant-producing bacteria," Applied Microbiology and Biotechnology, vol. 80, no. 6, pp. 1073–1085, 2008. · ·

100. N. Youssef, M. S. Elshahed, and M. J. McInerney, "Microbial processes in oil fields. Culprits, problems, and opportunities," Advances in Applied Microbiology, vol. 66, pp. 141–251, 2009.

101. D. R. Simpson, N. R. Natraj, M. J. McInerney, and K. E. Duncan, "Biosurfactant-producing Bacillusare present in produced brines from Oklahoma oil reservoirs with a wide range of salinities,"

Applied Microbiology and Biotechnology, vol. 91, no. 4, pp. 1083–1093, 2011. ··

102. D. G. Cooper, C. R. Macdonald, S. J. B. Duff, and N. Kosaric, "Enhanced production of surfactin fromBacillus subtilis by continuous product removal and metal cation additions," Applied and Environmental Microbiology, vol. 42, no. 3, pp. 408–412, 1981.

103. C. N. Mulligan, T. Y. K. Chow, and B. F. Gibbs, "Enhanced biosurfactant production by a mutantBacillus subtilis strain," Applied Microbiology and Biotechnology, vol. 31, no. 5-6, pp. 486–489, 1989.

104. G. Zheng and M. F. Slavik, "Isolation, partial purification and characterization of a bacteriocin produced by a newly isolated Bacillus subtilis strain," Letters in Applied Microbiology, vol. 28, no. 5, pp. 363–367, 1999. ··

105. H. Feitkenhauer, R. Müller, and H. Märkl, "Degradation of polycyclic aromatic hydrocarbons and long chain alkanes at 60–70°C by Thermus and Bacillus spp.," Biodegradation, vol. 14, no. 6, pp. 367–372, 2003. ··

106. J. Wang, T. Ma, J. Liu et al., "Isolation of functional bacteria guided by PCR-DGGE technology from high temperature petroleum reservoirs," Huan Jing Ke Xue, vol. 29, no. 2, pp. 462–468, 2008.

107. R. Hao, M. Lv, and A. Lu, "Biodegradation of crude oil in soil by Bacillus subtilis SB-1," Current Topics in Biotechnology, vol. 6, pp. 49–55, 2011.

108. G. Sanchez, A. Marin, L. Vierma, and T. P. Eugene, Isolation of Thermophilic Bacteria from a Venezuelan Oil Field, Developments in Petroleum Science, Elsevier, New York, NY, USA, 1993.

109. N. A. Sorkhoh, A. S. Ibrahim, M. A. Ghannoum, and S. S. Radwan, "High-temperature hydrocarbon degradation by Bacillus stearothermophilus from oil-polluted Kuwaiti desert," Applied Microbiology and Biotechnology, vol. 39, no. 1, pp. 123–126, 1993.

110. T. N. Nazina, D. S. Sokolova, A. A. Grigoryan et al., "Geobacillus jurassicus sp. nov., a new thermophilic bacterium isolated from a high-temperature petroleum reservoir, and the validation of the Geobacillus species," Systematic and Applied Microbiology, vol. 28, no. 1, pp. 43–53, 2005. ··

111. T. Kato, M. Haruki, T. Imanaka, M. Morikawa, and S. Kanaya, "Isolation and characterization of psychrotrophic bacteria from oil-reservoir water and oil sands," Applied Microbiology and Biotechnology, vol. 55, no. 6, pp. 794–800, 2001. · ·

112. M. L. Fardeau, B. Ollivier, B. K. C. Patel et al., "Thermotoga hypogea sp. nov., a Xylanolytic, thermophilic bacterium from an oil-producing well," International Journal of Systematic Bacteriology, vol. 47, no. 4, pp. 1013–1019, 1997.

113. G. Ravot, M. Magot, M. L. Fardeau et al., "Thermotoga elfii sp. nov., a novel thermophilic bacterium from an African oil-producing well," International Journal of Systematic Bacteriology, vol. 45, no. 2, pp. 308–314, 1995.

114. U.S. Environmental Protection Agency, National Contingency Plan, Product Schedule, 2013.

115. R. J. Portier and L. M. Basirico, Laboratory Screening of Commercial Bioremediation Agents for the Deepwater Horizon Spill Response, Department of Environmental Sciences, Louisiana State University, Baton Rouge, La, USA, 2011.

Chapter 8

Liquid Encapsulation Technology for Microelectromechanical Systems

Norihisa Miki[1]

[1]Department of Mechanical Engineering, Keio University, Hiyoshi, Kohoku-ku, Yokohama, Kanagawa, Japan

INTRODUCTION

Microelectromechanical systems (MEMS) have been extensively studied for over three decades, which has resulted in the prevalence of quite a few commercially available MEMS products in our daily lives, although they are too small to see. In the very beginning of the MEMS success story, people recognized the importance of packaging [1]. MEMS contain mechanical parts, and given their small sizes, they are severely affected by surrounding molecules. Therefore, MEMS are packaged under vacuum, at low pressure, or at least free from water molecules. Water molecules can bridge two separated parts and bring them into contact by the meniscus force, which may lead to permanent adhesion of the parts, known as stiction. This phenomenon must be averted, not only in the packaging, but also in the fabrication of parts. It is not an overstatement to say that researchers go to great lengths to keep their devices dry.

On the other hand, as MEMS technologies advance, a wide variety of applications are expected, some of which the MEMS must handle liquids. For example, drug delivery systems (DDS) that administer medicine to diseased parts at designated times can employ MEMS that are sufficiently small to be implanted and are capable of controlling discharge of the medicine [2-5]. In this application, the MEMS must contain medicine, which is in liquid form in many cases. In addition, MEMS can be used as a portable power source, referred to as power MEMS. Micro gas turbines and certain fuel cells require liquid fuel to generate chemical reactions [6-9]. Micro total analysis systems, or microTAS, are used to manipulate minute aqueous analytes and/or control microfluids to handle samples, such as cells and bacteria, for biochemical analysis [10-14].

Such useful characteristics of liquids are available to expand the design space for innovative MEMS devices. Functional liquids, such as magnetorheological fluid and electroconjugate liquids, can be used as micro pumps and actuators [15-16]. Other useful characteristics of liquids include deformability, incompressibility, and high dielectric constant. Hydraulic amplification can be achieved by exploiting the deformability and incompressibility of liquids [17-22]. In addition, highly dielectric liquids can enhance sensor sensitivity while maintaining flexibility [23].

While some applications allow such MEMS devices to bring the liquid from outside, encapsulation of liquid inside MEMS devices is mandatory in other applications. Liquid encapsulation technology can be used to manufacture innovative MEMS devices, such as completely spherical microlenses and hydraulic amplification mechanisms. Various liquid encapsulation technologies have been proposed to achieve these promising applications. The liquid species to be encapsulated and the application must be taken into consideration for the selection of appropriate encapsulation processes. This chapter reviews state-of-the-art liquid encapsulation technologies and their application to the manufacture of innovative MEMS devices that exploit the useful characteristics of the encapsulated liquids.

FILL AND SEAL TECHNIQUE

The most straight-forward technique to encapsulate liquids is to dispense liquid into a reservoir and then seal it with another substrate, as shown in Figure 1. The reservoirs can be easily manufactured using conventional MEMS fabrication technologies. Currently, commercially available dispensers are capable of dispensing a minute amount of liquid, as small as several nanoliters. Sealing, or bonding the substrates, is the most critical process.

In the MEMS field, bonding technologies have been widely explored for packaging and manufacturing three-dimensional structures [24-27]. The direct bonding of silicon wafers, anodic bonding of glass substrates, and thermocompression bonding using a metal thin film as an adhesive layer are examples of frequently used technologies. These technologies achieve strong bonding of substrates via covalent bonds; however, these processes have drawbacks when applied to the sealing of a liquid filled reservoir if they require high temperatures. High temperature processes in the order of several hundreds degrees Celsius may change the properties of the liquid to be encapsulated. For example, in drug delivery systems, it is necessary to maintain the medicinal effect of the encapsulated drug. Fuels for power MEMS applications should not be burnt before the device is completed. In addition, some liquids are volatile, which precludes not only high-temperature, but also vacuum processes.

Therefore, in this fill and seal approach, adhesive bonding shown in Figure 1 is the most appropriate technique. The adhesives employed include epoxy, UV curable resin including photoresist, and benzocyclobutene (BCB). Such adhesives solidify after either mixing with curing agents, exposure to UV irradiation, or thermal treatment at low temperatures.Either one-part or two-part epoxy resins can be used, because they do not require high temperature to promote solidification. The chemical reaction progresses with time, even at room temperature, and after solidification, the epoxy achieves strong bonds and is resistant to many chemicals.

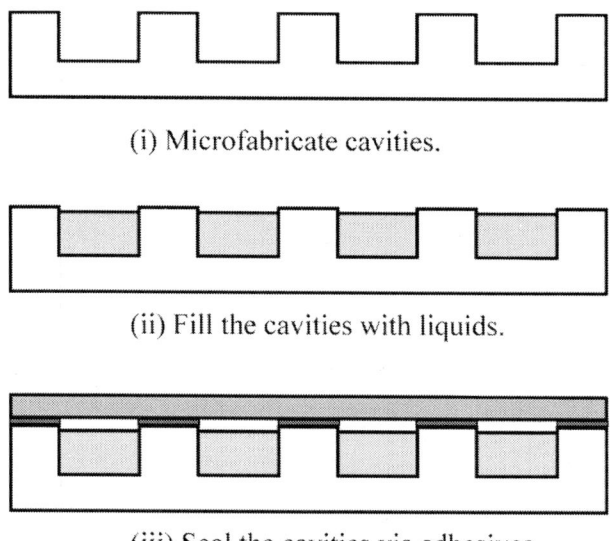

(i) Microfabricate cavities.

(ii) Fill the cavities with liquids.

(iii) Seal the cavities via adhesives.

Figure 1: Fill and seal approach.

While epoxy is a common adhesive, it is not compatible with conventional MEMS fabrication technology. To achieve reliable and reproducible bonding, the adhesives are preferably spin-coated, which allows the thickness to be controlled according to the spinning speed. In this regard, photoresist is a good candidate. Photoresists are compatible with MEMS fabrication technologies and can be spin-coated, and more importantly, knowledge of their use is well developed. Photoresist is coated on a substrate and then brought into contact with the pairing substrate either before or after the curing processes. A typical curing temperature is around 100 °C. When the contact is performed after curing, the bonding is achieved by a hot melt process at higher temperatures, although lower than 200 °C. Photoresist can be patterned using conventional photolithography to determine the bonding areas. The major drawback of using photoresists as adhesives is the weakness of the bond strength, i.e., they are not designed to function as adhesives. Adhesion between the photoresist and the substrates, as well as the mechanical strength of the photoresist, is designed to be sufficiently strong to survive photolithography processes. Therefore, the bonding may fail due to external forces, either at the interface or within the bulk.

BCB is a promising polymer adhesive that is photo-patternable and compatible with conventional photolithography processes. It can be spin-coated to thicknesses of 5-15 µm at spinning speeds of 1000-6000 rpm [28]. BCB has good chemical resistance, and the most significant advantage of this material is that it does not release any gases during the cure, which does not create pores in the material or contaminate the encapsulated liquid. BCB can be used to bond two substrates by thermocompression bonding. Compression at 230 °C has been attempted, which may limit the species of liquid to be encapsulated. However, BCB has been applied to seal sodium hypochlorite aqueous solution (NaOCl) for galvanic cells [28]. The paper discusses the BCB thickness and the bond quality determined by the geometry of the bonding areas.

UV curable resins do not require heat treatment, but only UV irradiation. If the MEMS devices are not UV sensitive and one substrate is transparent to UV light, then UV curable resin offers a strong bond after solidification with UV irradiation. Such bonding can even be conducted in liquids [21, 22, 29] and we have termed this the bonding-in-liquid technique (BiLT).

We have introduced sealing processes that employ polymer adhesives. However, the gas permeable nature of polymers may cause problems of contamination and vaporization of volatile liquids. For example, polydimethyl siloxane (PDMS), which is one of the most frequently used polymers in the fields of MEMS and microTAS, is permeable to gas. However, this permeability can be modified by the addition of different materials [30] or coating with airtight films [31]. Typical polymers are several orders of magnitude more permeable to gas than metals and ceramics [27]. Therefore, sealing with gold stud bumps has been proposed [32], where reservoirs are filled up with the liquids via microchannels and the inlets and outlets of the channels are then plugged with wire-bonding gold. Firstly, a gold ball is formed at the edge of the gold wire by electrical discharge. The ball is then pressed to the opening of the channel using ultrasound. The wire is then cut and the sealing is completed. Helium leak tests were conducted and hermetic sealing was verified using this technique when the hole diameters were less than 40 µm.

The inevitable drawback of the fill and seal approach is the filling rate; it is quite difficult to completely fill a reservoir with a liquid. This is acceptable for some applications, such as drug delivery and fuel supply for power MEMS devices. However, the performance of hydraulic displacement amplification mechanisms (HDAM) is deteriorated by the interfusion of compressible air. When liquids are used as components of sensors, contamination of gas or other liquids will lead to a loss of sensitivity. Therefore, liquid encapsulation techniques that enable complete filling of liquids are mandatory. The author's group developed BiLT, which is a fill and seal approach that enable complete filling [21, 22, 29].

BONDING-IN-LIQUID TECHNIQUE (BILT)

HDAMs require complete filling of the reservoir with an incompressible liquid, because gas is much more compressible than liquid. Figure 2 shows a package chamber that has openings at the top and bottom, where incompressible liquid is encapsulated with flexible polymer membranes. The top opening, which is determined by a metal plate, is smaller than the bottom opening; therefore, a small displacement applied to the bottom membrane is amplified at the top, according to the ratio of the openings. The application of HDAM is discussed in section 5.2. The key points in the fabrication of HDAMs are no interfusion of air bubbles and sealing with flexible polymer membranes. Complete filling can be achieved by the direct deposition of a thin film, which is detailed in the following section; however, this technique does not allow the use of flexible membranes. We have developed BiLT [29], which can be employed to overcome this problem.

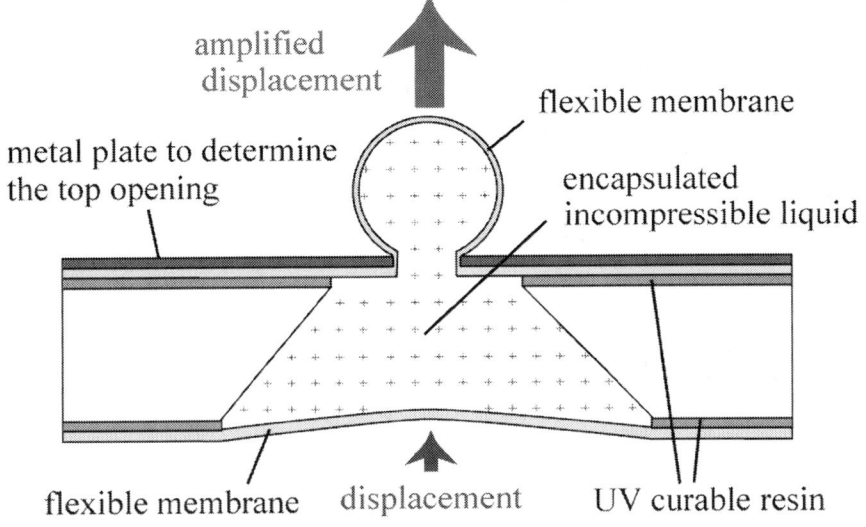

amplified
displacement

flexible membrane

metal plate to determine
the top opening

encapsulated
incompressible liquid

flexible membrane   displacement   UV curable resin

Figure 2: Schematic cross-sectional view of HDAM.

Rather than package the MEMS devices vacuum, we considered that if the encapsulating process was conducted in liquid, then the reservoir could be filled without the interfusion of air bubbles. However, one concern was how to successfully bond the membrane to the package chamber in a liquid environment. Therefore, it was decided to use a UV-curable resin (3164 Three Bond, Three Bond Co., Ltd.) that is solidified after UV irradiation, even in a liquid environment. This membrane achieves a tensile strength of 0.85 MPa when cured and the thickness of the resin can be controlled according to the spin-coating speed.

Figure 3 depicts the procedures employed in BiLT. Firstly, a UV resin is coated onto the bonding surface; however, it should be noted that the surface has many cavities for liquid encapsulation. Therefore, the UV resin is spin-coated onto a thick PDMS membrane and then transferred onto the bonding surface by soft contacting the PDMS membrane (Figure 3(a,d)). A sufficient amount of resin needs to be applied to the bonding surface to achieve a good bond, while excess resin may fall into and occupy the cavity during the bonding process. UV resin thicknesses of 80, 120, 160, and 200 μm were tested on PDMS membranes, which correspond to spin-coating speeds of 4000, 3000, 2000, and 1000 rpm, respectively. When silicon was used as the bonding substrate, the transferred thicknesses were 7.9, 8.1, 17, and 27 µm. In case of UV resin thicknesses of 17 and 27 µm, the excess resin flowed into the cavity.

Handling of a flexible thin membrane is not a trivial process. The membrane must be kept flat throughout the bonding process. Therefore, the membrane was spin-coated and cured on a glass substrate. The thickness of the PDMS membrane can be controlled according to the spin-coating speed. During the bonding process, the PDMS membrane must be peeled off the glass substrate. Therefore, the glass surface is coated in advance with a hydrophobic film (CYTOP M, H, Asahi Glass Corporation) to facilitate exfoliation.

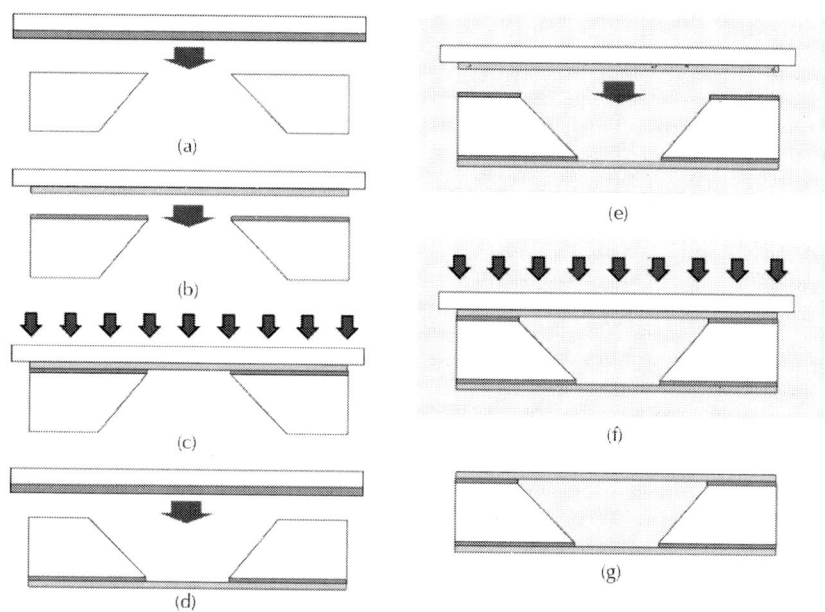

Figure 3: BiLT process. (a) UV curable resin is transferred onto the bonding surface. (b) A flexible membrane coated onto another substrate in advance is brought into contact with the bonding surface. The membrane is coated onto a hydrophobic layer to facilitate peeling of the membrane from the substrate. (c) UV light is irradiated to cure the UV resin. (d) A second UV curable resin is transferred onto the bonding surface. (e) A flexible membrane is brought into contact with the bonding surface in a liquid environment. (f) UV light is irradiated to cure the resin. (g) Liquid encapsulation without the interfusion of air bubbles or deformation of the membrane is achieved.

The substrate with cavities and the flexible membrane on the glass substrate are brought into contact in a liquid environment (Figure 3(e)). UV light is then irradiated onto the bonding surface through the glass substrate and flexible membrane to cure the UV curable resin (Figure 3(f)). Note that the substrate and membrane must be UV-transparent for this process. Figure 4 shows micrographs that confirm liquid (red-dyed deionized (DI) water) encapsulation was completed without the interfusion of air bubbles. Excess resin flowed into the cavities for UV resin thicknesses of 17 and 27 μm. No DI water was observed at the bonding interface. Encapsulation of glycerin was also attempted. Glycerin is non-volatile, so that the volume of the encapsulated glycerin did not change over a period of weeks even when encapsulated with a gas permeable PDMS membrane at ambient pressure and room temperature.

In HDAM, encapsulated liquids are sealed with flexible membranes at both the top and bottom sides of the package chamber. When the encapsulation/bonding process is conducted in air and not in liquid, the difference in the density of the air and liquid result in bowing of the membrane. Note that the membranes must be kept flat during the bonding processes of BiLT.

The bond strengths were investigated by conducting 180° peel tests on PDMS membranes and silicon substrates bonded using BiLT in DI water, glycerin, phosphate buffer solution (PBS), isopropyl alcohol (IPA), and acetone, and also in air as a reference. The silicon substrate used in the experiments did not contain bonding cavities. The bond strengths of the samples were measured as a function of time (1, 6, 24, 72, and 168 h) using a dynamic mechanical analyzer (RSAIII, TA Instruments). The test procedure involved one edge of the PDMS membrane being manually peeled from the silicon substrate and the unbonded area of the silicon substrate being clamped. The peeled PDMS membrane was then pulled in the direction parallel to the bonding interface at a speed of 3 mm/min until it peeled off, and the shear stress required to peel the PDMS membrane from the silicon substrate was measured. The results are shown in Figure 5. The bonding resin was dissolved in both IPA and acetone solution, and thus bonding was unsuccessful when conducted in these solutions, while the bonding strengths of the other samples were comparable. The bond strengths increased with time, most likely due to continuing chemical reaction of the UV-curable resin over time. The bonding strengths after 1 week were more than 4 times greater than those obtained after 6 h. Peel tests conducted within 72 h of bonding revealed failure of the resin, while failure occurred at the interface between the resin and PDMS membrane when measured after 1 week. This indicates that failure occurred within the resin until the resin was sufficiently cured and this is why the bonding strengths in air, DI water and PBS were similar; 1 week after bonding, the bonding strengths achieved by bonding in air and using BiLT were comparable.

The developed BiLT enables complete liquid filling with various membranes. Many species of liquids can be encapsulated using BiLT, unless the liquids dissolve the UV resin. This feature is crucial in manufacturing HDAM and sensors, which will be introduced in section 5. Complete filling can be achieved by direct deposition of a thin film, as introduced in section 4; however, this process can only be used to encapsulate non-volatile liquids, and the type of sealing membrane is also limited. The major drawback of this technology is that the substrate must be UV-transparent and the device should not contain UV-sensitive materials. For example, dye-sensitized photovoltaic cells, which are employed as transparent solar cells and optical sensors, require the encapsulation of electrolytes. However, BiLT cannot be used for encapsulation, because the cells have dyes that degrade after being exposed to UV.

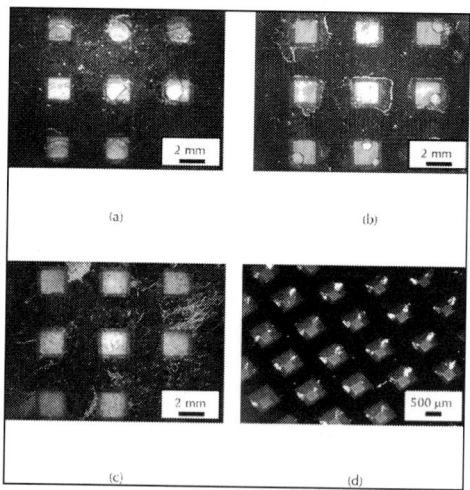

Figure 4: Bonding results for resin thicknesses of (a) 27, (b) 17 and (c) 8.1 μm. Red-dyed water was encapsulated into the cavities. The trans-

parent parts in the cavities are excess resin. When a certain amount of resin was used, excess UV resin or air was found in the cavities, as shown in (d).

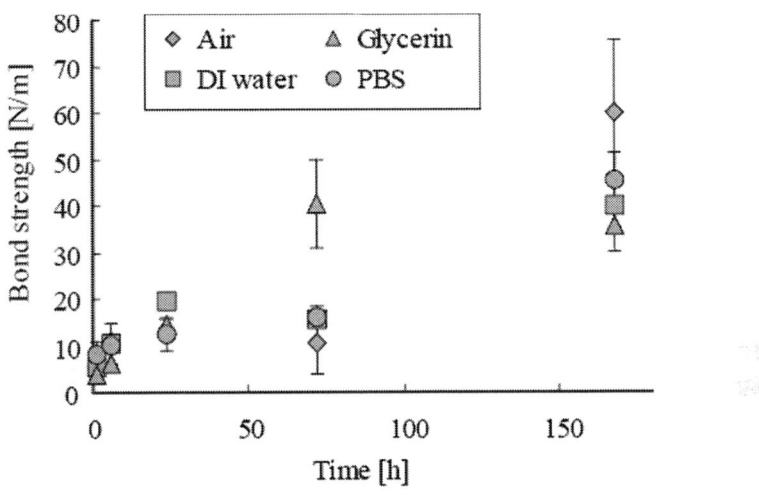

Figure 5: Bond strength after BiLT in different media as a function of time.

LIQUID ENCAPSULATION BY DIRECT DEPOSITION OF A THIN FILM

Thin film deposition onto a solid is typically conducted under vacuum. Some liquids, such as silicone oil and ionic liquids have extremely low vapor pressure and do not evaporate under vacuum. A thin film of metal or polymer can be directly deposited onto such low-vapor-pressure liquids. A thin silver film was deposited onto an ionic liquid to manufacture a mirror for a space telescope [33].

Parylene, or poly(para-xylylene), is widely used in fields of MEMS and microTAS due to favorable characteristics, such as transparency, mechanical strength (3.2 GPa), biocompatibility, gas sealing efficacy, and has the ability to be conformally coated using chemical vapor deposition [34-36]. Parylene has been used to form a microspring with a low spring constant [37] and as a substrate and/or a protective layer for the manufacture of microelectrodes [38]. In addition, the surfaces of PDMS microchannels have been coated with parylene to protect against protein adsorption [39].

The typical pressure for parylene deposition is several pascals. Therefore, liquids with vapor pressures less than this can remain in the liquid phase during the parylene deposition process. Binh-Khiem et al. proposed parylene on liquid deposition (POLD), where parylene is directly deposited onto a low-vapor-pressure liquid, such as silicone oil [34-36]. A feature of liquids is that sufficiently small droplets can have perfectly spherical shape due to surface tension; however, when the droplets are not small, the shape is deformed by gravity. Spherical droplets can be used as lenses. The focal length of a lens is in the same order as the lens diameter; therefore, small lenses enable compact optical systems. POLD can be used to form spherical microlens arrays; transparent liquid can be directly deposited onto silicone oil droplets. A film is formed at the droplet surface, so that no air is included in the lens, i.e., a filling rate of 100% is achieved. Parylene is more flexible than metals; therefore, by integrating electrodes beneath the liquid lens coated with parylene, the lens can be deformed to vary the focal length according to the voltage applied to the electrodes [34].

The direct deposition on liquid approach enables perfect liquid encapsulation with good reproducibility. The useful characteristics of parylene or specific metals can be exploited; however, the disadvantage is that the liquids to be encapsulated must be non-volatile and the type of sealing material is limited. For example, this approach cannot be applied to HDAM, because it is preferable to seal the liquid with flexible membranes.

APPLICATIONS OF LIQUID ENCAPSULATION TECHNOLOGY

Drug Delivery

In DDS, the release of medicine is designed so that the drug efficacy is high. MEMS-based DDS are expected to convey medicine to the vicinity of diseased sites by exploiting the small MEMS size and drug release occurs when appropriate. Santini et al. proposed the controlled release of drugs from cavities that were sealed with a thin gold layer [2]. The thin gold layer is used as an anode and an electrochemical reaction occurs, with subsequent dissolution of the gold film and release of the encapsulated medicine. Drug delivery is thus controlled by electrical control of the reaction.

In this application, the fill and seal approach is employed to maintain the quality of the medicine as a priority. The package chambers are prepared with a thin gold layer as the bottom surface. Drugs are dispensed into the chambers and then sealed with a water-proof membrane using epoxy adhesive. Details of this process can be found in the literature [2].

Hydraulic Amplification

Some MEMS applications require large displacement of several tens of micrometers. High-flow-rate microvalves are one such application. To satisfy the requirements, hydraulic amplification has been reported that utilizes incompressible fluid in a microchamber with an input surface that is larger than the output surface [17-20]. Another example is a tactile display [21, 22]; Figure 6 shows an illustration of an array of MEMS actuators that mechanically deform the fingertip and stimulate tactile receptors. These receptors typically require skin deformation of several tens of micrometers, which could be facilitated by hydraulic amplification. Tactile displays are developed to offer a new approach to human-machine interface for virtual reality applications and interactive devices such as pointing devices or game controllers, and also to support visually impaired persons. In the microvalve applications, a working fluid can be used as an incompressible fluid for hydraulic amplification. However, tactile displays cannot afford to have an external fluidic system to drive a working fluid; therefore, complete encapsulation of the liquid is necessary. In addition, the sealing membranes must be flexible. HDAMs have been successfully developed by encapsulating incompressible and non-volatile glycerin in microchambers with flexible and largely deformable PDMS membranes (mixture of DC 3145 CLEAR and RTV thinner, Dow Corning Toray Inc.) via UV curable resin using BiLT. Figure 2 shows a schematic of a HDAM [21, 22].

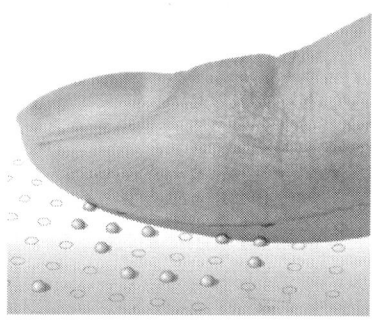

Figure 6: Tactile display that employs an array of MEMS actuators.

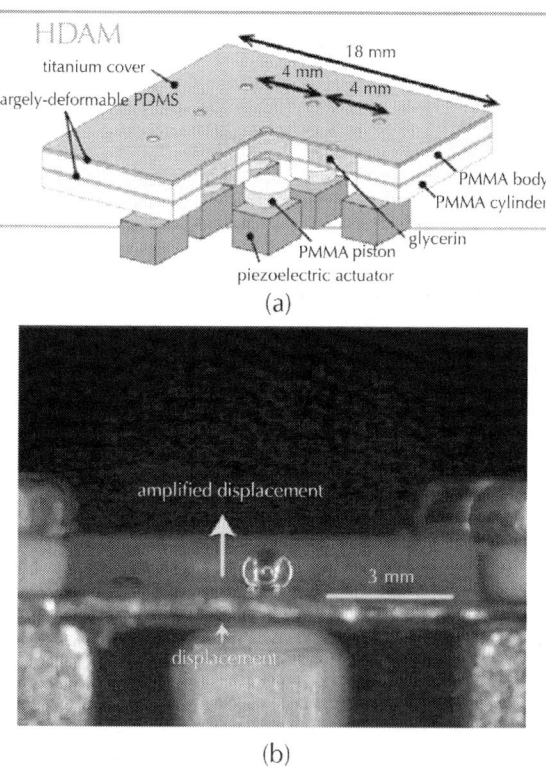

Figure 7: Large displacement MEMS actuator array that consists of HDAM with glycerin-filled cavities prepared using BiLT, and piezoelectric actuators. (a) Schematic image of HDAM, and (b) micrograph showing the large deformation of a PDMS membrane.

The HDAM shown in Figure 7(b) was combined with piezoelectric actuators and applied to develop large-displacement MEMS-actuators, with a particular aim to application in MEMS-based tactile displays [21, 22, and 40]. When applied to a vibrational Braille code display, it was experimentally verified that the large-displacement HDAM could display Braille codes more efficiently than a static display. This is because both fast and slow adaptive tactile receptors could be used to detect the displayed patterns when individual cells were vibrated at several tens of hertz [40]. When the actuation of the large displacement MEMS actuators was controlled both spatially and temporary, different surface textures, such as rough and smooth, could be displayed.

Flexible Capacitive Sensor

Highly sensitive pressure sensors are expected to be applied in humanoid robots and medical instruments to detect tactile sensation, which would enable safe physical interaction with the environment, including human contact. MEMS-based capacitive sensors that have simple structures composed of electrodes and a dielectric component have been widely studied, due to good compatibility with MEMS fabrication technologies. Capacitive sensors require not only high sensitivity, but also flexibility to detect the pressure applied to curved surfaces. Silicon-based MEMS capacitive sensors have been developed; however, silicon is brittle, which makes it difficult for the sensors to conform to a curved surface. Therefore, polymer-based flexible sensors have been proposed and demonstrated. Polymer-based flexible sensors are typically used to maintain flexibility with air as the dielectric; however, air has a relatively low dielectric constant. A solid dielectric may enhance the sensitivity, but impairs the flexibility of the sensor. Therefore, a polymer-based capacitive sensor that uses a dielectric liquid has been proposed, as depicted in Figure 8 [23]. DI water and glycerin have high relative dielectric constants of approximately 80.4 and 47; therefore, the proposed sensor with such liquids can have high sensitivity while maintaining flexibility. The capacitance of the electrodes increases when pressure is applied to the device. PDMS is used as a structural material in this device. An escape reservoir is designed to allow an incompressible liquid, such as DI water and glycerin, to move from the cavity between the electrodes when pressure is applied to the sensor, which allows the flexible sensors to deform and vary the capacitance. The proposed microsensor has been fabricated, and both high sensitivity and flexibility have been experimentally demonstrated.

Dye-Sensitized Photovoltaic Cell

Dye-sensitized photovoltaic cells are currently attracting widespread scientific and technological interest as a high efficiency, low-cost, and transparent alternative to inorganic solar cells. Figure 9shows a schematic illustration of the structure and operation principle of the dye-sensitized photovoltaic device. The cell consists of two electrodes and an encapsulated liquid electrolyte that contains iodide and triiodide ions. The cathode is a highly porous nanocrystalline semiconductive titanium dioxide (TiO_2) layer, in many cases consisting of TiO_2 nanoparticles, deposited on a transparent electrically conductive glass. TiO_2 absorbs only UV light; therefore, dye is adsorbed onto the TiO_2 layer to utilize the light with a wider range of wavelength. The counter electrode (anode) is a transparent electrically conductive glass with a platinum catalyst. The device is transparent and is colored according to the dye employed.

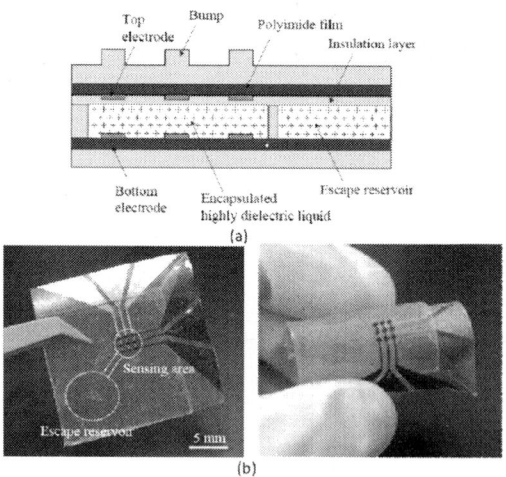

Figure 8: a) Cross-sectional view of a capacitive sensor with encapsulated liquid dielectric to enhance the sensitivity while maintaining flexibility. (b) Fabricated capacitive sensor.

When light passes through the electrically conductive transparent glass electrode, the dye molecules are excited and transfer an electron to the semiconducting TiO_2 layer via electron injection. The electron is then transported through the TiO_2 layer and collected by the conductive layer on the glass. The mediator (I-/I3-) undergoes oxidation and regeneration in the electrolyte. Electrons lost by the dye molecules to the TiO_2 layer are replaced by electrons from the iodide and triiodide ions in the electrolyte, thereby generating iodine or triiodide, which in turn obtains electrons at the counter electrode, culminating in a current flow through the external electrical load. This is the mechanism for the conversion of light energy received by the device to electricity [41]. This device has an interesting feature in that it reacts strongly to light that enters through the TiO_2 layer.

The dye-sensitized photovoltaic cell has been conventionally studied as a solar cell, where miniaturization was not considered. However, when the cells are microfabricated and arrayed, they can be used as a transparent optical sensor. Shigeoka et al. proposed to microfabricate a transparent optical sensor on eyeglasses, which could detect the pupil position by detecting reflection from the eye, as shown in Figure 10 [42]. The light reflected from the pupil is considered to be smaller than that from the white. The sensor reacts strongly to light from the TiO_2 electrode side, i.e., when the TiO_2 layer electrode is faced towards the eyes, it detects only the light reflected from the pupil and white of the eye, without being affected by the light incident on the device from the environment.

Figure 11 shows a schematic of the processes used for fabrication of this device. The most critical part is encapsulation of the electrolyte. The conductive layer (ITO) is firstly patterned on the glass substrate using photolithography. TiO_2 nanoparticles are patterned on the cathode using a lift off process. The device is subsequently annealed in air at 450 °C for 60 min and then dipped in a ruthenium-containing dye solution for 60 min to ensure the dye is adsorbed onto the TiO_2 nanoparticles. The two glass substrates are bonded via a hot melt film and application of 600-800 kPa at 100 °C. Lastly, the liquid electrolyte is flowed from the inlet hole into a channel between the two electrodes, and then the inlet and outlet holes are covered by end seals. The dyes used are UV sensitive; therefore, BiLT was not applicable to this liquid encapsulation process and the fill and seal approach was used instead. However, the filling rate of the electrolyte was quite high and no interfusion of air between the electrodes was observed. An array of the dye-sensitized photovoltaic devices successfully detected the pupil position. The line-of-sight (LOS) was successfully deduced [42, 43] from the obtained pupil position and the front image of the subject, acquired using a CCD camera on the eyeglasses.

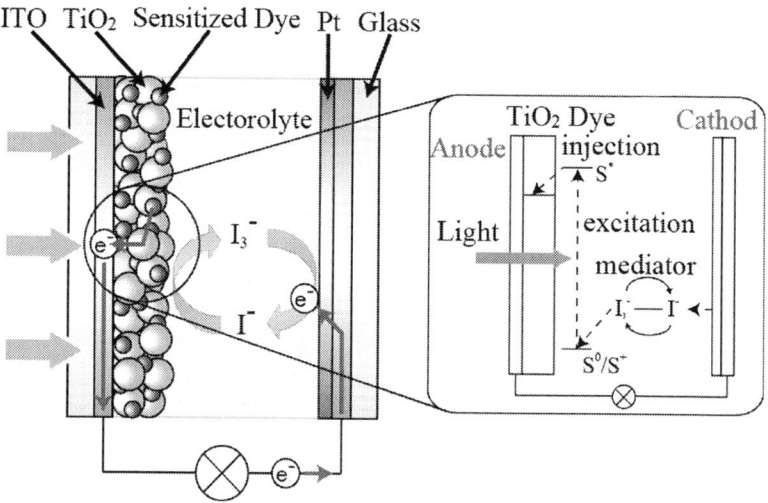

Figure 9: Structure and operation principle of the dye-sensitized photovoltaic device.

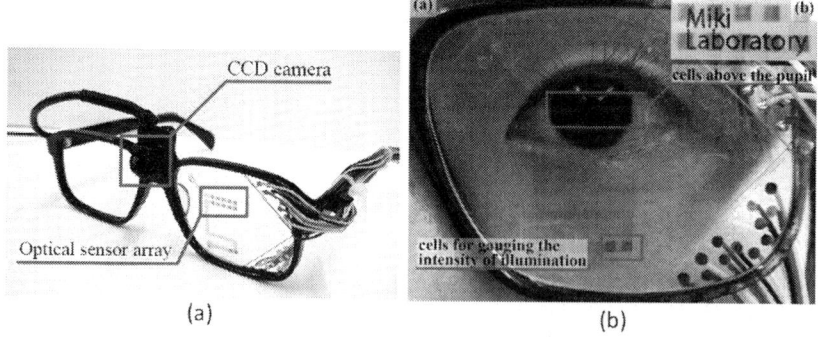

Figure 10: a) Array of dye sensitized photovoltaic cells patterned onto eyeglasses to detect the pupil position. The electrolyte was encapsulated between the electrodes. (b) Photograph of the sensor when worn by a subject.

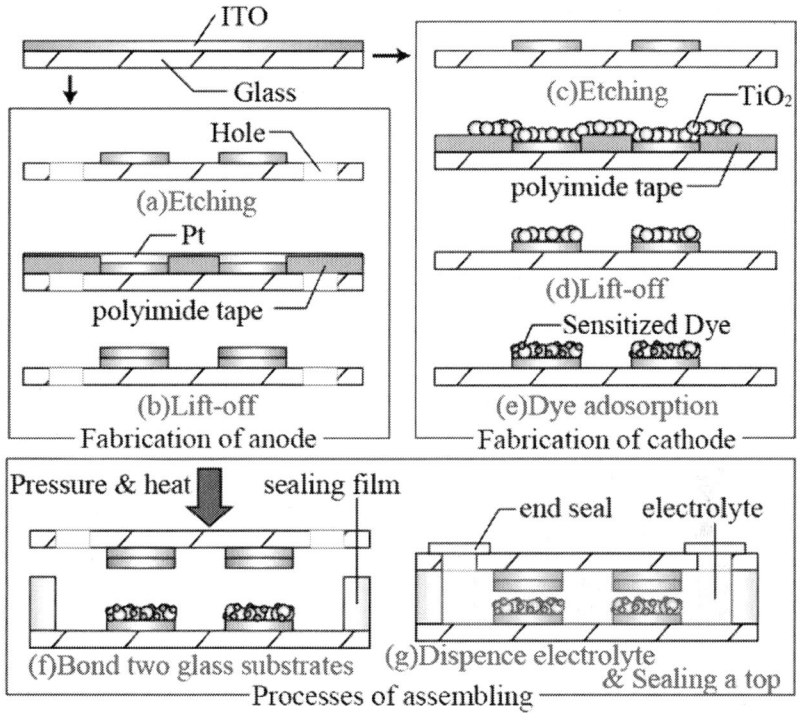

Figure 11: Fabrication process to produce an array of dye-sensitized photo-voltaic devices. The fill and seal approach was employed to encapsulate the electrolyte (g).

CONCLUSIONS

This chapter has reviewed liquid encapsulation technologies and their applications to the manufacture of innovative MEMS devices that exploit the useful characteristics of liquid; liquid is a deformable and liquid droplet form a perfectly spherical shape by surface tension. Other liquids have high relative dielectric constants. Liquids can be used as drugs for DDS and fuels for power MEMS. Appropriate liquid encapsulation technologies must be selected according to the liquid to be encapsulated. The fill and seal approach, bonding-in-liquid technique, and direct deposition of a thin film were discussed in this chapter, all of which have both advantages and disadvantages.

The use of liquid in MEMS packaging is quite a new technology. The author is convinced that more and more liquid encapsulation technologies will be developed and contribute to the further development of innovative liquid-encapsulating MEMS devices.

ACKNOWLEDGEMENTS

This work was supported by a Grant-in-Aid for Young Scientists (B) (21760202) from the Ministry of Education, Culture, Sports, Science and Technology (MEXT) of Japan, the Strategic Information and Communications R&D Promotion Programme (SCOPE) (092103005) of the Japan Ministry of Internal Affairs and Communications (MIC), and the Information Environment and Humans research area of PRESTO (Precursory Research for Embryonic Science and Technology) from the Japan Science and Technology Agency (JST).

REFERENCES

1.   S. D Senturia, Microsystem Design. Kluwer Academic Publishers; 2001.

2.   J. T Santini, M. J Cima, R Langer, A controlled-release microchip. Nature 1999397335338.

3.   J. T Santini, A. C Richards, R Scheidt, M. J Cima, R Langer, Microchips as controlled drug-delivery devices. Angewandte Chemie International Edition 20003923962407.

4.   Y Li, R. S Shawgo, B Tyler, P. T Henderson, J. S Vogel, A Roesnberg, P. B Storm, R Langer, H Brem, M. J Cima, In vivo release from a drug delivery MEMS device. Journal of Controlled Release 2004100211219.

5.   Richards Grayson ACShawgo RS, Li Y, Cima MJ. Electronic MEMS for triggered delivery. Advanced Drug Delivery Reviews 200456173184.

6.   A. H Epstein, S. D Senturia, Macro power from micro machinery. Science 1997.

7.   N Miki, C. J Teo, L. C Ho, X Zhang, Enhancement of rotordynamic performance of high-speed micro-rotors for power MEMS

applications by precision deep reactive ion etching. Sensors and Actuators A: Physical 2003104263267.

8. J. D Holladay, E. O Jones, M Phelps, J. L Hu, Microfuel processor for use in a miniature power supply. Journal of Power Sources 2002.

9. L. R Arana, S. B Schaevitz, A. J Franz, M. A Schmidt, K. F Jensen, Journal of Microelectromechanical Systems 2003125600612.

10. A Manz, N Graver, H. M Widmer, Miniaturized total chemical-analysis systems- a novel concept for chemical sensing. Sensors and Actuators B: Chemical; 1244248.

11. S Shoji, Fluids for sensor systems. Topics in Current Chemistry 1998194163168.

12. K Huikko, R Kostiainen, T Kotiaho, Introduction to micro-analytical systems: bioanalytical and pharmaceutical applications. European journal of pharmaceutical sciences 200320149171.

13. H Ota, T Kodama, N Miki, Rapid formation of size-controlled three-dimensional hetero-cell aggregates using micro-rotation flow for spheroid study. Biomicrofluidics 2011.

14. Y Gu, N Miki, Multilayered microfilter using a nanoporous PES membrane and applicable as the dialyzer of a wearable artificial kidney. Journal of Micromechanics and Microengineering 2009.

15. S Yokota, K Kawamura, K Takemura, K Edamura, High-integration micromotor using electro-conjugate fluid (ECF). Journal of Robotics and Mechatronics 2005172142148.

16. A Yamaguchi, K Takemura, S Yokota, K Edamura, A robot hand using electro-conjugate fluid. Sensors and Actuators A: Physical 2011170139146.

17. D. C Roberts, L Hanqing, J. L Steyn, O Yaglioglu, S. M Spearing, M. A Schmidt, N. W Hagood, A piezoelectric microvalve for compact high-frequency, high-differential pressure hydraulic micropumping systems. Journal of Microelectromechanical Systems 20031218192.

18. H Kim, K Najafi, Electrostatic hydraulic three-way gas microvalve for high-pressure applications. Proceedings of the 12th International Conference on Miniaturized Systems for Chemistry and Life Sciences, MicroTAS 20081216October 2008San Diego, USA.

19.   H Kim, K Najafi, An electrically-driven large-deflection high-force, micro piston hydraulic actuator array for large-scale microfluidic systems. Proceedings of 22th IEEE International Conference on Micro Electro Mechanical Systems, MEMS 20092529January 2009Sorrento, Italy.

20.   X Wu, S. H Kim, C. H Ji, M. G Allen, A piezoelectrically driven high flow rate axial polymer microvalve with solid hydraulic amplification. Proceedings of 21st IEEE International Conference on Micro Electro Mechanical Systems, MEMS 20081317January 2008Tuscon, USA.

21.   X Arouette, Y Matsumoto, T Ninomiya, Y Okayama, N Miki, Dynamic characteristics of a hydraulic amplification mechanism for large displacement actuators systems. Sensors 20101029462956.

22.   T Ninomiya, Y Okayama, Y Matsumoto, X Arouette, K Osawa, N Miki, MEMS-based hydraulic displacement amplification mechanism with completely encapsulated liquid. Sensors and Actuators A: Physical; 166277282

23.   Y Hotta, Y Zhang, N Miki, A flexible capacitive sensor with encapsulated liquids as dielectric. Micromachines 20123137149.

24.   Q. Y Tong, U Gosele, Semiconductor wafer bonding science and technology: John Wiley & Sons, Inc.; 1999.

25.   N Miki, Wafer bonding techniques for MEMS. Sensor Letters 200534263273.

26.   N Miki, S. M Spearing, Effect of nanoscale surface roughness on the bonding energy of direct-bonded silicon wafers. Journal of Applied Physics 20039468006806.

27.   F Niklaus, G Stemme, J. Q Lu, R. J Gutmann, Adhesive wafer bonding. Journal of Applied Physics 2006.

28.   J Dlutowski, C. J Biver, W Wang, S Knighton, J Bumgarner, L Langebrake, W Moreno, A. M Cardenas-valencia, The development of BCB-sealed galvanic cells. Case study: aluminum-platinum cells activated with sodium hypochlorite electrolyte solution. Journal of Micromechanics and Microengineering 20071717371745.

29.   Y Okayama, K Nakahara, A Xavier, T Ninomiyia, Y Matsumoto, A Hotta, M Omiya, N Miki, Characterization of a bonding-in-liquid

technique for liquid encapsulation into MEMS devices. Journal of Micromechanics and Microengineering 2010.

30. Y Zhang, M Ishida, Y Kazoe, Y Sato, N Miki, Water vapour permeability control of PDMS by the dispersion of collagen poweder. TEEE: Transactions on Electrical and Electronic Engineering 200943442449.

31. S Sawano, K Naka, A Werber, H Zappe, S Konishi, Sealing method of PDMS as elastic material for MEMS. Proceedings of 21st IEEE International Conference on Micro Electro Mechanical Systems, MEMS 20081317January 2008Tuscon, USA.

32. M Antelius, A. C Fischer, F Niklaus, G Stemme, N Roxhed, Hermetic integration of liquids using high-speed stud bump bonding for cavity sealing at the wafer level. Journal of Micromechanics and Microengineering 2012.

33. E. F Borra, O Seddiki, R Angel, D Eisenstein, P Hickson, K. R Seddon, S. P Worden, Deposition of metal films on an ionic liquid as a basis for a lunar telescope. Nature 20074477147979981.

34. N Binh-khiem, K Matsumoto, I Shimoyama, Polymer thin film deposited on liquid for varifocal encapsulated liquid lenses. Applied Physics Letters 2008

35. N Binh-khiem, K Matsumoto, I Shimoyama, Tensile film stress of parylene deposited on liquid. Langmuir 201026241877118775.

36. S Takamatsu, H Takano, N Binh-khiem, T Takahata, E Iwase, K Matsumoto, I Shimoayma, Liquid-phase packaging of a glucose oxidase solution with parylene direct encapsulation and an ultraviolet curing adhesive cover for glucose sensors. Sensors 201010658885898.

37. Y Suzuki, Y. C Tai, Micromachines high-aspect-ratio paryelene spring and its application to low-frequency accelerometers. Journal of Microelectromechanical Systems 200615513641370.

38. S Takeuchi, D Ziegler, Y Yoshida, K Mabuchi, T Suzuki, A parylene flexible neural probe integrated with micro fluidic channels. Lab on a Chip 20055519523.

39. H Sasaki, H Onoe, H Osaki, R Kawano, S Takeuchi, Parylene-coating in PDMS microfluidic channels prevents the absorption of fluorescent dyes. Sensors and Actuators B: Chemical 20101501478482.

40. J Watanabe, H Ishikawa, X Arouette, Y Matsumoto, N Miki, Demonstration of vibrational Braille code display using large displacement micro-electro-mechanical system actuators. Japanese Journal of Applied Physics 2012FL11.

41. O Regan, B Gratzel, M. A low cost, high-efficieny solar cell based on dye-sensitizd colloidal TiO2 films. Nature 1991.

42. T Shigeoka, T Muro, T Ninomiya, N Miki, Wearable pupil position detection system utilizing dye-sensitized photovoltaic devise. Sensors and Actuators A: Physical 2008.

43. A Oikawa, N Miki, MEMS-based eyeglass type wearable line-of-sight detection system. Proceedings of 2011 IEEE International Conference on Robotics and Automation, ICRA 2011913May 2011Shanghai, China.

Chapter 9

Mysteries behind the Low Salinity Water Injection Technique

Emad Waleed Al-Shalabi, Kamy Sepehrnoori, and
Gary Pope

Department of Petroleum and Geosystems Engineering, The University
of Texas at Austin, Austin, TX 78712, USA

ABSTRACT

Low salinity water injection (LSWI) is gaining popularity as an improved oil recovery technique in both secondary and tertiary injection modes. The objective of this paper is to investigate the main mechanisms behind the LSWI effect on oil recovery from carbonates through history-matching of a recently published coreflood. This paper includes a description of the seawater cycle match and two proposed methods to history-match the LSWI cycles using the UTCHEM simulator. The sensitivity of residual oil saturation, capillary pressure curve, and relative permeability parameters (endpoints and Corey's exponents) on LSWI is evaluated in this work. Results showed that wettability alteration is still believed to be the main contributor to the

LSWI effect on oil recovery in carbonates through successfully history matching both oil recovery and pressure drop data. Moreover, tuning residual oil saturation and relative permeability parameters including endpoints and exponents is essential for a good data match. Also, the incremental oil recovery obtained by LSWI is mainly controlled by oil relative permeability parameters rather than water relative permeability parameters. The findings of this paper help to gain more insight into this uncertain IOR technique and propose a mechanistic model for oil recovery predictions.

INTRODUCTION

Oil recovery from carbonate rocks is a challenge due to the high fracture density and the rock wettability state which ranges from mixed-wet to oil-wet. One of the recently recommended improved oil recovery (IOR) techniques is low salinity water injection (LSWI), which is believed to shift the wettability state of the rock towards more water-wet state. The LSWI technique has several advantages including high efficiency in displacing light to medium gravity crude oils, ease of injection into oil-bearing formations, availability and affordability of water, and lower capital and operating costs compared to other IOR methods, which leads to favorable economics. Other names proposed in the literature for the same mechanism are LoSal, Smart Waterflood, and Advanced Ion Management. The LSWI effect on oil recovery from carbonates was shown both at laboratory scale and to a limited extent at field scale. Although most researchers believe that wettability alteration is the main mechanism for the LSWI on oil recovery from carbonates, there are others who believe in the presence of other contributing mechanisms. Therefore, work is still progressing to understand the chemical interactions in crude oil-brine-rock (COBR) in the porous media.

The LSWI effect on oil recovery from carbonate is not well addressed compared to sandstone rocks due to the previous thoughts of relating wettability alteration by low salinity water to the presence of clay, which is not the case in carbonate rocks. Nevertheless, the effect of LSWI on oil recovery from carbonate rocks was investigated at laboratory scale using both spontaneous imbibition and coreflooding studies.

Høgnesen et al. [1] concluded from their spontaneous imbibition experiments on reservoir limestone cores that increasing sulfate ion concentration at high temperature leads to oil recovery increase due to the role of sulfate ion as a wettability modifying agent for carbonate rocks from mixed-wet to water-wet state. Webb et al. [2] investigated the effect of sulfate on oil recovery from North Sea carbonate core samples through spontaneous imbibition experiments. They reported that seawater has the ability to alter wettability of the carbonate system to more water-wet state compared to sulfate-free water. Zhang et al. [3] studied wettability alteration of North Sea chalk reservoirs in Ekofisk field showing the effect of adding calcium and/or magnesium ions at various temperatures. They concluded that wettability alteration occurs if the imbibing water contains either Ca^{2+} and SO_4^{2-} or Mg^{2+} and SO_4^{2-}. Moreover, Strand et al. [4] observed 15% increase in oil recovery from limestone cores when seawater was imbibed compared to seawater free of sulfate. Increased oil recovery using low salinity water injection in limestone formations was noticed by Fjelde [5].

Bagci et al. [6] reported high oil recovery of 35.5% of OOIP from their corefloods by using 2 wt% KCl on limestone cores and high pH effluent brine due to ions exchange reactions with the clay present in the rock. They considered wettability alteration as the reason behind recovering more oil without further explanation. Yousef et al. [7] investigated the applicability of low salinity water injection (Smart Waterflood) on carbonate rocks for improving oil recovery by using different dilutions of seawater. The results of coreflooding tests showed increasing oil recovery with stepwise dilution of seawater upon which 18% incremental oil recovery was achieved due to tertiary water injection. Coreflooding experiments on both dolomite cores from West Texas and limestone cores from the Middle East were performed by Gupta et al. [8]. Experiments showed incremental 5–9% OOIP recovery from both dolomite and limestone cores as a result of adding sulfate ions. For limestone cores, 7–9% OOIP was obtained due to reducing hardness of the injected water, not the total dissolved solids. Another interesting finding is the 15% and 20% OOIP by using borate (BO_3^{3-}) and phosphate (PO_4^{3-}) as modified ions, respectively. In a later work, Yousef et al. [9] demonstrated that wettability alteration is the reason behind LSWI through NMR, contact angle measurement, and zeta potential studies. The results showed that wettability alteration occurs through changing the surface charge from the zeta potential measurements and dissolution of $CaSO_4$ from NMR tests.

Al-Harrasi et al. [10] provided direct evidence of low salinity water flooding effect on oil recovery from Omani carbonate rocks through spontaneous and coreflooding experiments. Wettability alteration was referred to as the reason for low salinity water injection with negligible reduction in interfacial tension. Also, they reported that low salinity water injection by lowering the ionic strength has more pronounced effect on oil recovery from oil-bearing zone cores compared to hardening the injected water. Nevertheless, the case is just the opposite for Stevns Klint outcrop chalk cores because they are more responsive to hardening of the injected low salinity water as was reported by Romanuka et al. [11] through their spontaneous imbibition experiments.

Extensive research was performed by Austad and coworkers [12–14] which showed the possibility of wettability alteration and enhancing oil recovery from carbonate rocks by modifying the ionic composition in the injected water. Wettability alteration is the main and most acceptable mechanism for the incremental oil recovery obtained from carbonate rocks using LSWI. Wettability alteration in carbonate rocks using smart water can be achieved by injecting water containing SO_4^{2-} and either Ca^{2+} or Mg^{2+} or both of them in the presence of high temperatures (>90°C). It was proposed that, with increasing temperature, the affinity of sulfate to chalk rock surface increases and sulfate adsorption occurs. At the same time, Ca^{2+} adsorption increases as well, as the initial positive charge of the rock decreases. Hence, more excess Ca^{2+} ions are present close to the surface which reacts with the carboxylic material and releases some of them. Moreover, with increasing the temperature Mg^{2+} becomes more active, Ca^{2+} substitution by Mg^{2+} occurs, and sulfate becomes less active as it reacts with Mg^{2+}. Otherwise, $CaSO_4$ precipitation occurs which causes injection problems [13].

The first ever LoSal application in carbonate reservoirs was reported by Yousef et al. [15]. Two single-well chemical tracer tests (SWCTT) were applied in an upper jurassic carbonate reservoir using a diluted version of Qurayyah seawater. The tests resulted in about 7 saturation units' reduction in the residual oil beyond conventional seawater injection. The results obtained matched their previous experimental work which is encouraging to plan a multiwell demonstration pilot.

Only a few modeling works for carbonate rocks have been performed so far due to various reasons. These reasons include the complex chemical interaction in rock-oil-brine and the heterogeneity of carbonate rocks, which complicate oil recovery predications by LSWI. Also, the uncertainty in the controlling mechanism and the clash in some of the published experimental results shifted the focus on experimental work rather than modeling work. This paper includes history-matching of Chandrasekhar and Mohanty's [16] experimental work using the UTCHEM simulator, which is a 3D multiphase flow, transport, and chemical flooding simulator for black oil, developed at The University of Texas at Austin.

EXPERIMENTAL DATA

Chandrasekhar and Mohanty [16] conducted several vertical corefloods to investigate the low salinity water injection effect on oil recovery from Middle Eastern carbonate core plugs. The coreflood of our interest is the one with different injected seawater dilutions. Heterogeneous carbonate core plug was used with average porosity and liquid permeability of 26.4% and 7.59 mD, respectively. The coreflood was conducted at reservoir temperature of 248°F and atmospheric pressure at an injection rate of 1 ft/day. At each injection cycle, the injection rate was increased to 10 ft/day to make sure that the oil maximum recovery was reached. The brines used were field water of 179,726 ppm and different diluted versions of seawater of 43,619 ppm by weight. Rock properties and fluid properties are shown in Tables 1, 2, and 3. The oil recovery and pressure drop data are shown in Figures 1 and 2. Here an increase in oil recovery is observed with stepwise dilution of the seawater by tertiary water injection. More details about the experimental work are described elsewhere [16]. The UTCHEM software was used to history-match both oil recovery and pressure drop data. The pressure drop data of each injection cycle is matched just for the low injection rate (1 ft/day). The second injection cycle is an exception, where the pressure drop is matched for both injection rates (1 and 10 ft/day) because of the incremental oil recovery obtained with increasing the rate at 10 ft/day in this cycle.

Table 1: Reservoir core properties (Chandrasekhar and Mohanty, 2013 [16])

Pore Volume	15.8 cc
Porosity	0.264
Permeability	7.59 mD
Diameter	3.79 cm
Cross Sectional Area	11.28 cm²
Length	5.3 cm
S_{wi}	0.3181
S_{oi}	0.6819

Table 2: Field oil sample properties (Chandrasekhar and Mohanty, 2013 [16])

Oil properties	
Viscosity (cp) at 248°F	1.05
Density (g/cc) at 248°F	0.618
API (degrees)	40

Table 3: Properties of various dilutions of seawater (248°F) (Chandrasekhar and Mohanty, 2013 [16])

Water propertiesat 248°F		
Injected water salinity (ppm)	Water density (g/cc)	Water viscosity (cp)
43619	0.9760	0.260
21809.5	0.9595	0.246
4361.9	0.9464	0.234
2180.95	0.9448	0.233

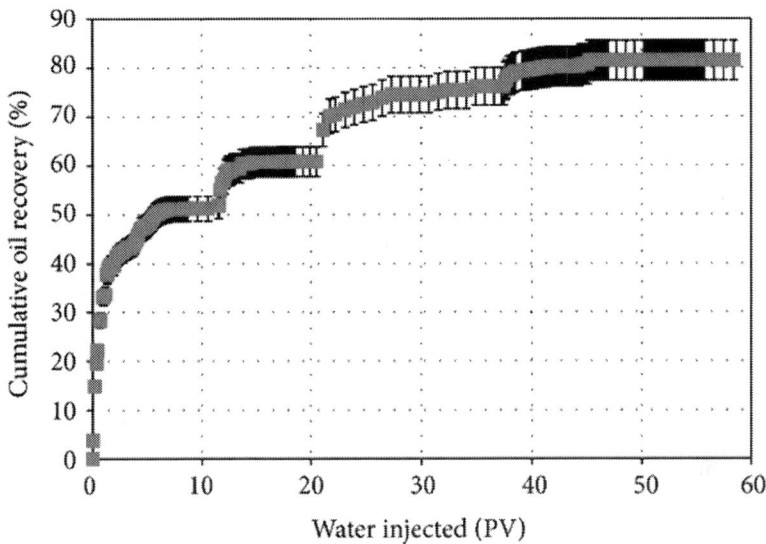

Figure 1: Cumulative oil recovery curve for the coreflood conducted [16].

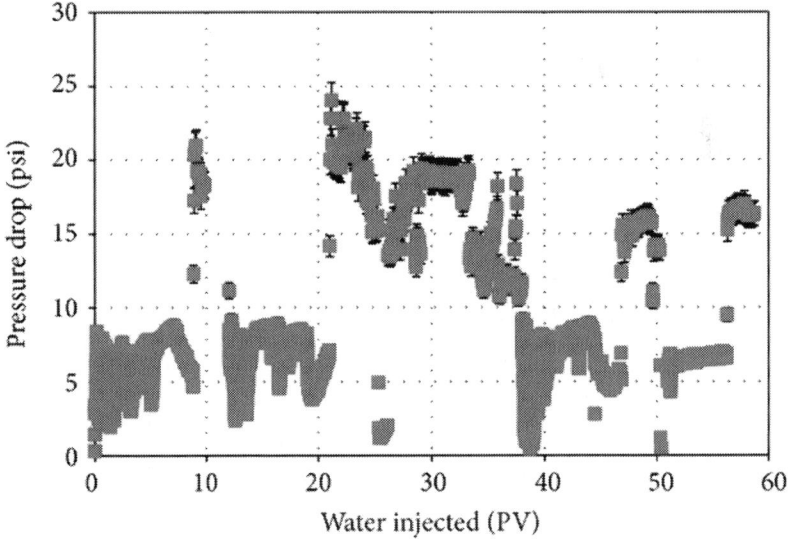

Figure 2: Pressure drop data for the coreflood conducted [16].

SIMULATION DATA

This section includes both simulation model description and experimental data analysis to obtain the simulation inputs needed.

Simulation Model Description

A 2D Cartesian grid system was used with 10 × 1 × 10 grid blocks to simulate the heterogeneous core plug for the coreflood. The decision on using a heterogeneous model is discussed later in the seawater cycle match section. The heterogeneous model was considered by generating permeability distribution with an arithmetic mean of 7.59 mD and Dykstra Parson's coefficient of 0.85. A spherical variogram and a log normal permeability distribution were used. The y-direction correlation length was assumed to be similar to the x-direction; however, a high z-direction correlation length compared to the x-direction was chosen, which generated vertical layers of different permeabilities (Figure 3). The simulation model has two horizontal wells, injector at the bottom and producer at the top. Table 4 shows length, width, and height dimensions for the grid blocks. We assumed a negligible capillary end effect.

Table 4: Heterogeneous core model data

Parameter	Value	Comments
Number of grid blocks	100	2D (10 × 1 × 10)
Grid block sizes (ΔI, ΔJ, ΔK), m	x-direction: 1–10, Δx is 0.0033588 m y-direction: 1-1, Δy is 0.033588 m z-direction: 1–10, Δz is 0.0053 m	Constant grid size in the x-, y-, and z-direction.
Composite core model dimensions, m	0.033588 m × 0.033588 m × 0.053 m	Length × width × height

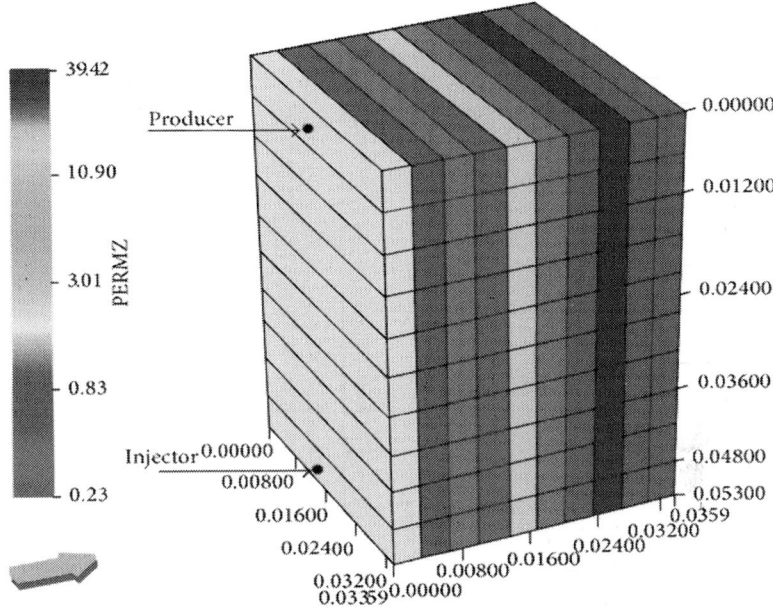

Figure 3: Simulation model used in different runs with heterogeneous permeability.

Experimental Data Analysis

This section includes a description of pressure drop data analysis and application of JBN method to find relative permeability curves for the seawater injected cycle.

Pressure Drop Data Analysis

As was previously mentioned, the analysis of the pressure drop curve (Figure 2) is performed for the 1 ft/day injection rate, except for the second cycle where the injection rate of 10 ft/day resulted in additional oil recovery. The water endpoint relative permeability for each cycle at 1 ft/day and even for the second cycle at 10 ft/day can be calculated using Darcy's law and stabilized average pressure drop value. The oil endpoint relative permeability was provided experimentally for the seawater cycle ($K_{ro}^* = 0.203$). Table 5 summarizes the obtained endpoint relative permeabilities for water and oil. Table 5 shows endpoint permeability calculation for the second cycle including both low salinity water injection and trapping number effects. We can see a slight decrease in the water endpoint relative permeability values for the LSWI effect in all injection cycles, which indicates the presence of wettability alteration by LSWI.

Table 5: Endpoint relative permeability data analysis

Oil viscosity	1.05	cP	Oil-water IFT	30	Dynes/cm
Composite core length	5.3	cm			
Cross-sectional area	11.28	Cm²			
Injection rate (main)	0.045	cc/min	$K_{o@Swirr}$	1.54	mD
Absolute brine permeability	7.59	mD			
Injection cycle	Water viscosity (cP)	Pressure drop (psi)	K_{rw}^{*}	S_{or}	K_{ro}^{*}
First	0.26	7.20	0.025	0.329	0.203
Second 1 (LSWI effect)	0.246	7.00	0.024	0.267	
Second 2 (trapping number effect)	0.246	18.90	0.089	0.163	
Third	0.234	7.00	0.023	0.127	
Fourth	0.233	7.20	0.022	0.127	

JBN Method

Johnson, Bossler, and Naumann (JBN) method was applied to find relative permeability curves for the seawater cycle. The data obtained are shown in Figure 4; Corey's model was fitted to find relative permeability endpoints and exponents for the seawater cycle. Moreover, the analysis was taken a step further to calculate the fractional flow curve, upon which both mobility ratio and gravity effects are considered. The fractional flow in this case is defined as

$$f_w = \frac{S^{n_w} M^o \left[1 - N_g^o (1 - S)^{n_o} \sin(\alpha) \right]}{S^{n_w} M^o + (1 - S)^{n_o}}.$$

$$(1)$$

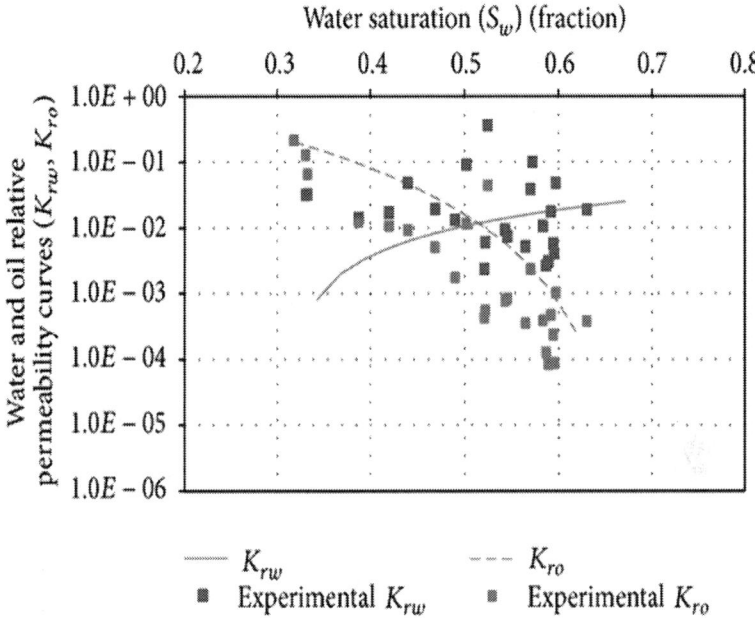

Figure 4: Relative permeability data analysis using the JBN method (first cycle).

The relative permeability data calculated from JBN method was fitted to an exponential function that was used later to calculate the fractional flow curve through mobility ratio calculations (Figure 5). The equation of relative permeability ratio as function of saturation is given by

$$\frac{K_{ro}}{K_{rw}} = ae^{-bS_w}.$$

(2)

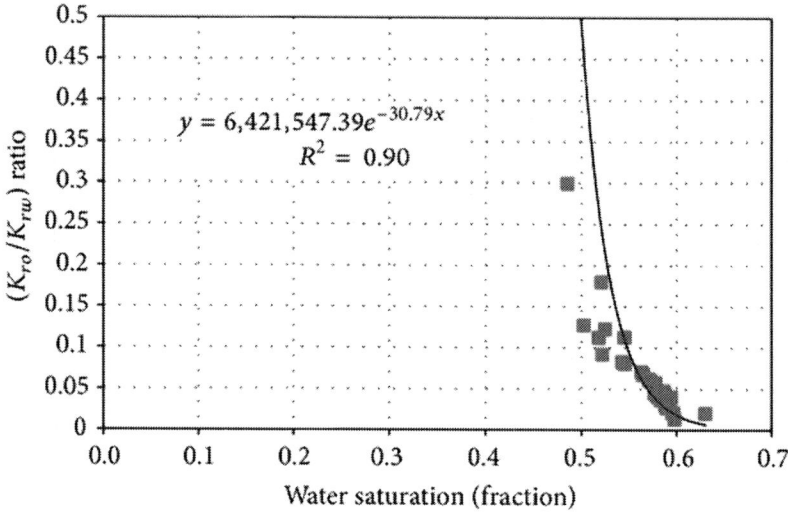

Figure 5: Relative permeability ratio versus water saturation (first cycle).

The latter fractional flow curve (experimental) was matched with the analytical solution of fraction flow, where the main matching parameters are n_w of 1.3 and n_o of 3.5 (Figure 6). The water and oil endpoint relative permeability values were obtained using pressure drop curve analysis from Table 5. The coreflood was conducted vertically which makes sin( ) equal to 1. The endpoint mobility ratio used is 0.47 and the endpoint gravity number is 0.01. The definitions of endpoint mobility ratio (M^o) and endpoint gravity number (N_g^o) are given by

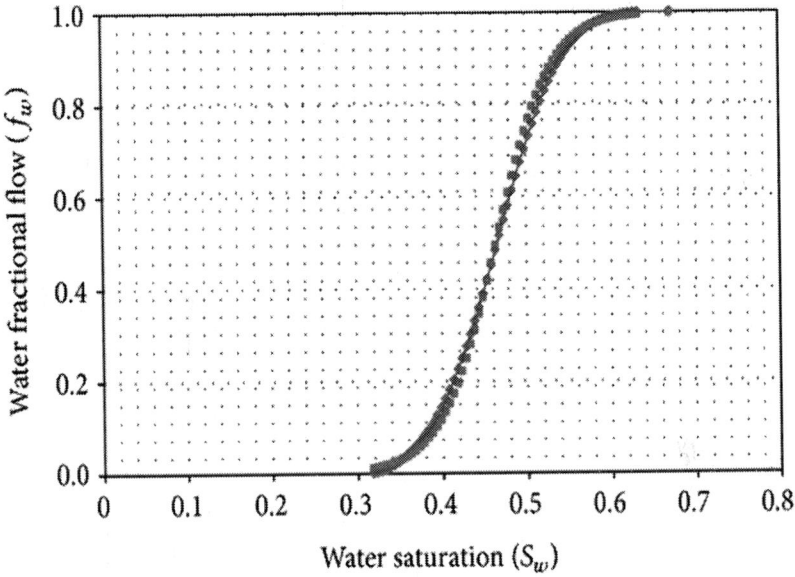

Figure 6: Fractional flow curve history-match (first cycle).

Nevertheless, there is a mismatch between the Buckley Leverett analytical solution compared to the experimental data provided by Chandrasekhar and Mohanty [16] (Figure 7). The main difference between the analytical solution and the experimental data is the heterogeneity effect which is not taken into account in analytical solution added to the capillary pressure effect. Both of these effects can be considered using the UTCHEM simulator to history-match the data.

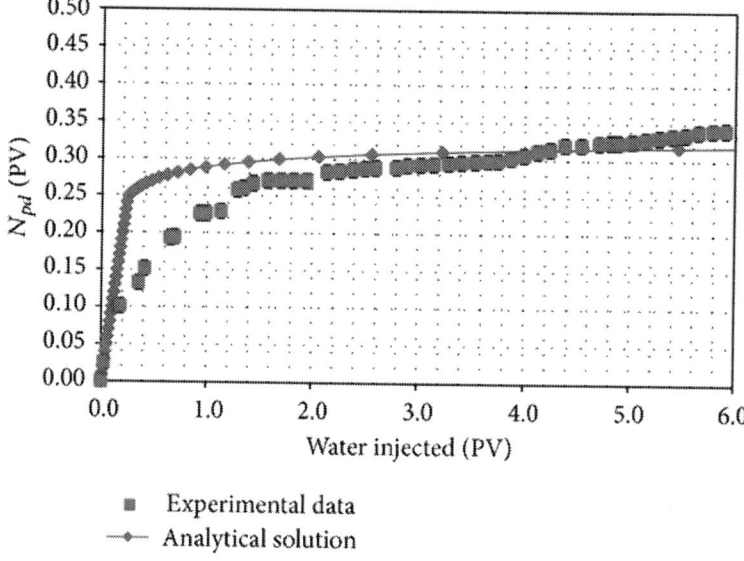

Figure 7: Buckley Leverett analytical solution (first cycle).

RESULTS AND DISCUSSION

This section covers seawater cycle history-matching and methods used for the wettability alteration effect matching for different dilutions of seawater injected cycles.

Seawater Cycle Match

The final set of relative permeability curves for the seawater cycle as a result of pressure drop analysis and JBN method is shown in Figure 8. The figure shows a weakly oil-wet rock, where K_{ro}^* is 0.203, K_{rw}^* is 0.025, n_o is 3.5, n_w is 1.3, and the intersection point is at about 0.5 water saturation. The heterogeneity effect was taken into consideration by applying a Dykstra Parson coefficient of 0.85 along with correlation lengths which resulted in vertical layers of different permeabilities as was previously shown in Figure 3. Moreover, the capillary pressure contribution was considered by applying Brooks-Corey model for imbibition capillary pressure for mixed-wet rocks as follows.

(i)   Water-wet part of capillary pressure curve ($S < S^*$) is

$$P_{c12} = CPC_1 \sqrt{\frac{\phi}{k}} \left(\frac{S_w^* - S_w}{S_w^* - S_{wr}} \right)^{EPC_1} ;$$

(4)

(ii) Oil-wet part of capillary pressure curve ($S > S^*$) is

$$P_{c12} = CPC_2 \sqrt{\frac{\phi}{k}} \left(\frac{S_w - S_w^*}{1 - S_{or} - S_w^*} \right)^{EPC_2} ,$$

(5)

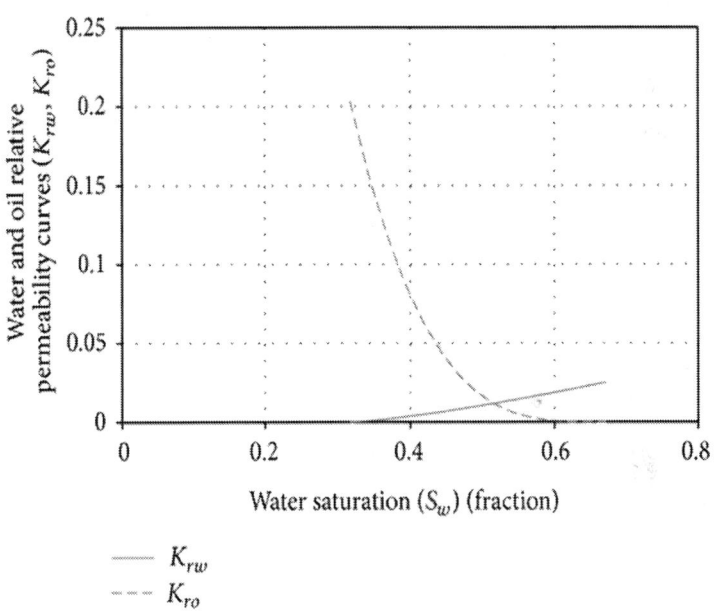

Figure 8: Relative permeability curves (first cycle).

where CPC is a parameter related to the maximum capillary pressure, EPC is capillary pressure exponent, and S^* is the water saturation at zero capillary pressure value. The capillary pressure curve used in matching oil recovery and pressure drop data along with summary of capillary pressure parameters and relative permeability parameters is presented in Figure 9 and Table 6, respectively. History-matching of oil recovery and pressure drop data showed that the CPC_2 parameter vcontrols the ultimate oil recovery value; however, CPC_1 controls the initial hump of the oil recovery and the pressure drop data match. It can be seen clearly that the capillary pressure does not contribute much to data history-matching. Hence, the capillary pressure is neglected for history-matching the successive dilutions of seawater injection.

Table 6: Summary of relative permeability and capillary pressure parameters (seawater cycle)

Seawater cycle match parameters			
Relative permeability parameters			
k_{rw}^*	0.025	n_w	1.3
k_{ro}^*	0.203	n_o	3.5
Capillary pressure parameters			
CPC_w	2		2
CPC_o	−2		2
S^*	0.5		

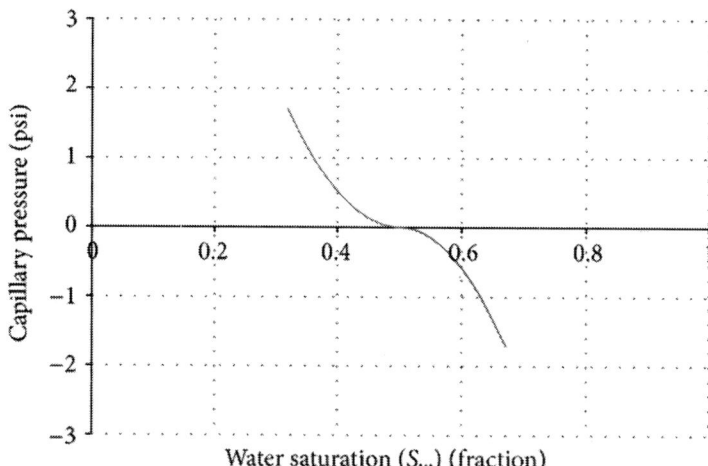

Figure 9: Capillary pressure curve (first cycle).

The results of history-matching of oil recovery and pressure drop data are depicted in Figures 10 and 11, respectively. In the latter figures, two curves are presented for the homogeneous 1D model with an average permeability and heterogeneous models. The history-matching shows the importance of heterogeneity incorporation to match reasonably the oil recovery and pressure drop curves.

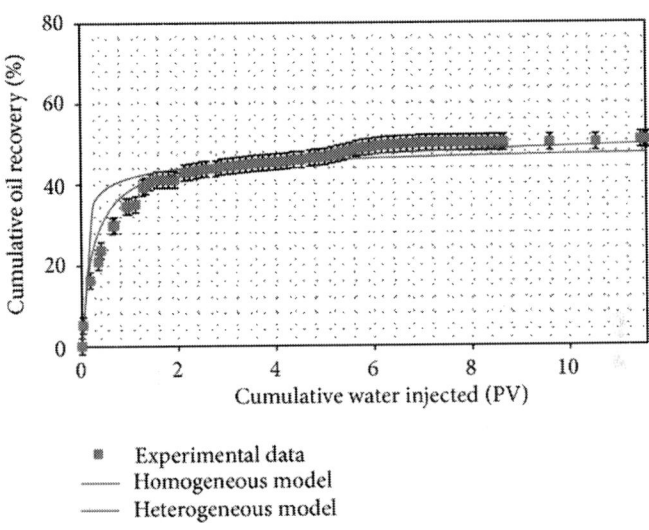

Figure 10: Oil recovery match for seawater cycle.

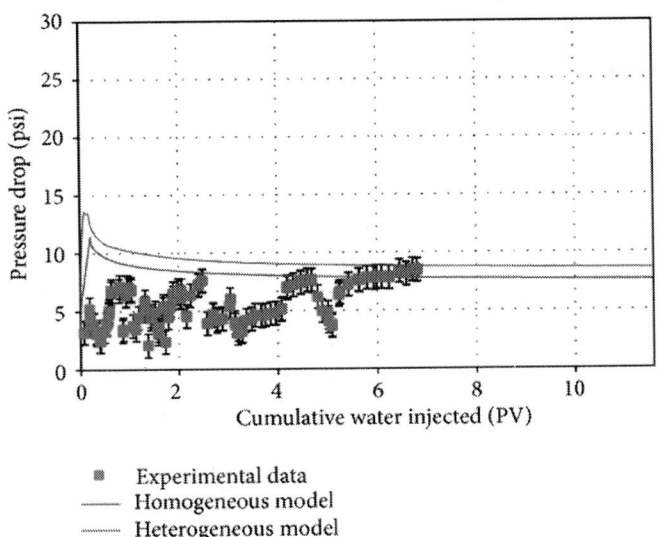

Figure 11: Pressure drop match for seawater cycle.

Dilutions of Seawater Injected Cycles Match

This section includes history-matching of the LSWI cycles of Chandrasekhar and Mohanty [16] coreflood using two proposed methods.

First Method

In this method, seawater cycle's relative permeability parameters are used for the different dilution cycles, while only changing the residual oil saturation for each cycle based on the reported values. As expected, history-matching of data is not possible using this method which validates the necessity of tuning relative permeability parameters for LSWI cycles because S_{or} contribution by itself is not enough. This is supported by the findings of oil recovery and pressure drop history-matching curves using this approach (Figures 12 and13). It is worth mentioning that the jump in the second cycle of pressure drop curve (Figure 13) is due to the trapping number effect as the injection rate was increased to 10 ft/day without changing the relative permeability parameters.

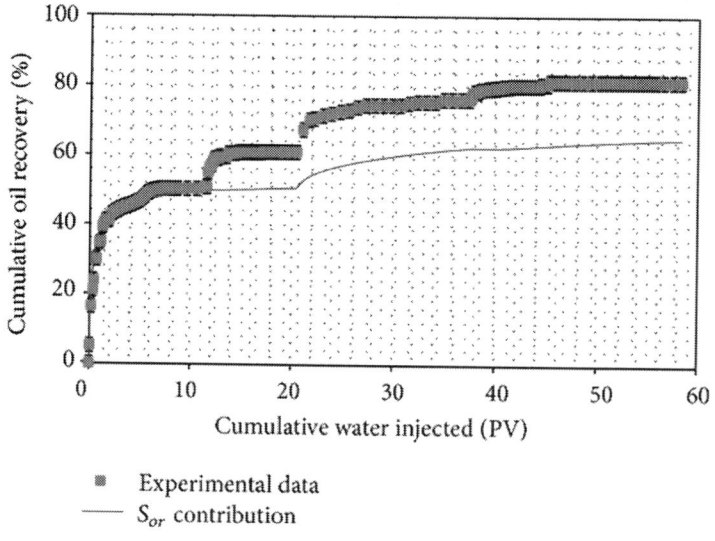

Experimental data
—— S_{or} contribution

Figure 12: Cumulative oil recovery match using the first method.

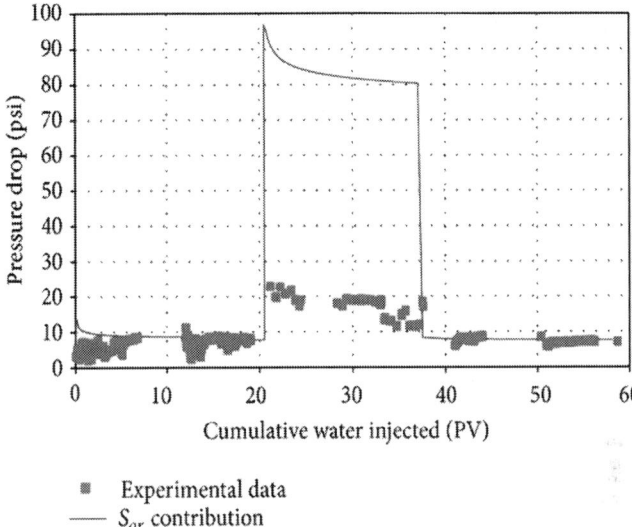

Figure 13: Overall pressure drop match using the first method.

Second Method

This method includes three approaches: changing Corey's exponents only (first approach), changing endpoint relative permeabilities only (second approach), and changing both Corey's exponents and endpoint relative permeabilities (third approach). The first two approaches were not successful in history-matching pressure drop and oil recovery data. The third approach of the second method is applied on Chandrasekhar and Mohanty [16] coreflood by tuning relative permeability parameters, including endpoints and Corey's exponents to match the data in each cycle starting with the second cycle. The K_r and S_{or} contributions curve for the LSWI effect on the second cycle is shown in Figure 14.

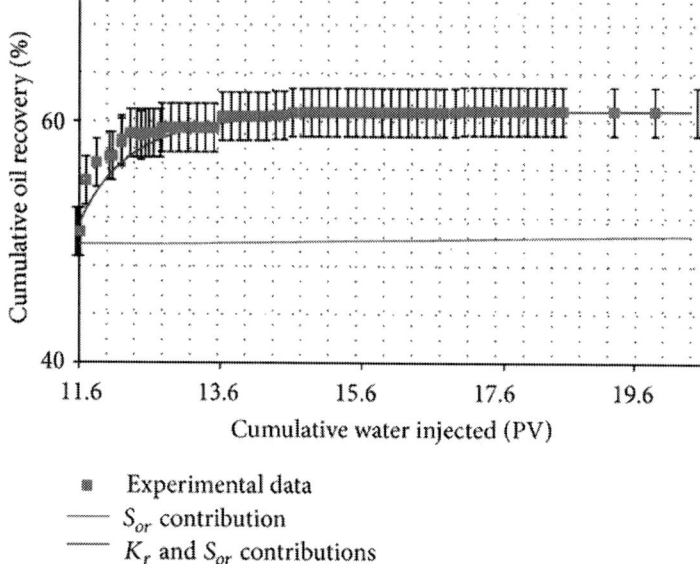

Experimental data
— S_{or} contribution
— K_r and S_{or} contributions

Figure 14: LSWI effect on second cycle oil recovery match using the second method (third approach).

The trapping number effect on the second cycle is considered using the capillary desaturation curve (CDC). The relation for adjusting residual oil saturation as a function of trapping number was proposed by Pope et al. [17] as follows:

$$S_{lr} = S_{lr}^{high} + \frac{S_{lr}^{low} - S_{lr}^{high}}{1 + T_l N_{T_l}^{\tau}}, \quad \text{for } l = 1, \ldots, np.$$

(6)

Figure 15 shows the modeled CDC curve for the second cycle where the experimental trapping number calculated for the injection rate of 10 ft/day is matched using S_{lr}^{low} of 0.267, S_{lr}^{high} of zero, of 0.82, and T_l parameter of 650,000. The detailed calculations of the CDC curve are listed in Table 7. The effect of trapping number on relative permeability parameters was also considered using Delshad et al.'s [18] proposed model as follows:

$$k_{rl}^o = k_{rl}^{o^{low}} + \frac{S_{l'r}^{low} - S_{l'r}}{S_{l'r}^{low} - S_{l'r}^{high}} \left(k_{rl}^{o^{high}} - k_{rl}^{o^{low}} \right),$$

$$\text{for } l, l' = 1, \ldots, np,$$

$$n_l = n_l^{low} + \frac{S_{l'r}^{low} - S_{l'r}}{S_{l'r}^{low} - S_{l'r}^{high}} \left(n_l^{high} - n_l^{low} \right)$$

$$\text{for } l, l' = 1, \ldots, np. \tag{7}$$

Table 7: CDC curve parameters

S_{or}(high)	0.000
S_{or}(low)	0.267
T_{22}(parameter)	650,000
Tau (N^Texponent)	0.82
N^T	S_{or}
1.00E − 11	0.267
1.00E − 11	0.267
1.00E − 10	0.266
1.00E − 09	0.260
5.00E − 09	0.242
1.00E − 08	0.226
1.00E − 07	0.122
1.00E − 06	0.030
5.00E − 06	0.009
1.00E − 05	0.005
1.00E − 06	0.001
1.00E − 04	0.000

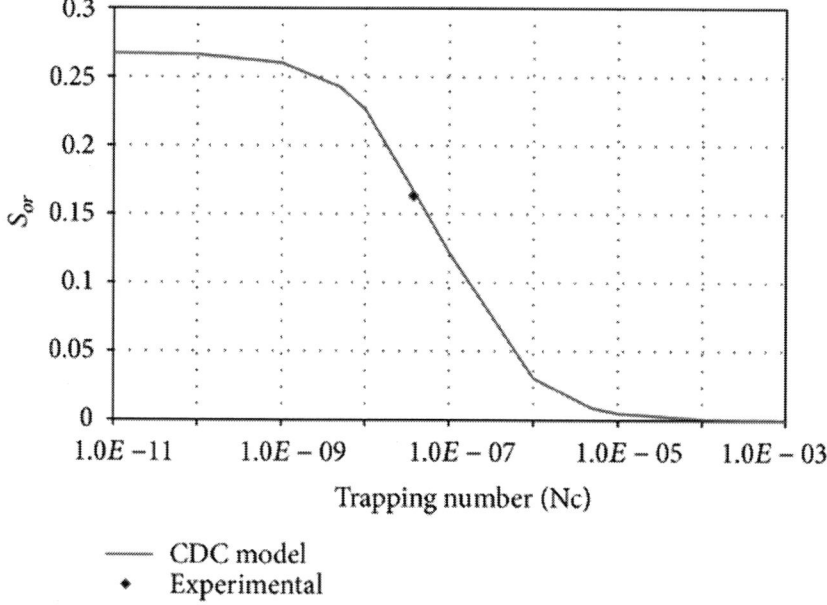

Figure 15: Modeled CDC curve for the coreflooding experiment.

In the previous equations, the words "high" and "low" in the superscripts indicate the value of the parameter at high and low trapping numbers, respectively. The values at high trapping number are usually assumed and the values at low trapping number can be considered as the values obtained through history-matching the effect of LSWI on the second cycle. It is worth mentioning that $K_{rl}^{o^{high}}$ was assumed to be 0.2 due to the low water endpoint relative permeability of initial seawater cycle (0.025). Table 8 and Figure 16 show two sets of relative permeability curves before and after exceeding the critical trapping number. The oil recovery and pressure drop match for trapping number effect on the second cycle are shown in Figures 17 and 18, respectively.

Table 8: Relative permeability parameters before and after exceeding critical N^T

Second cycle matching parameters trapping number effect			
Below N_T^*(critical)		**Exceeding N_T^*(critical)**	
n_w	1.7	n_w	1.43
n_o	1.55	N_o	1.55
K_{rw}^*	0.024	K_{rw}^*	0.089
K_{ro}^*	0.83	K_{ro}^*	0.83
S_{or}	0.267	S_{or}	0.163
S_{wirr}	0.3181	S_{wirr}	0.3181

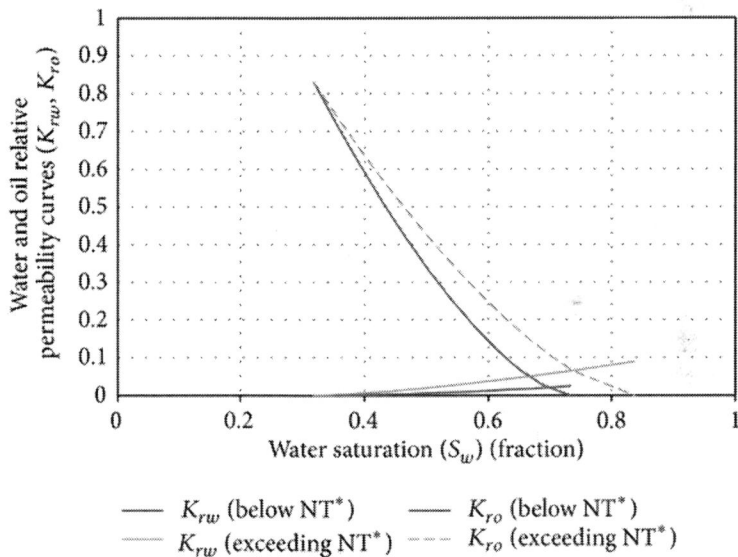

Figure 16: Relative permeability curves before and after exceeding critical N^T (second cycle-trapping number).

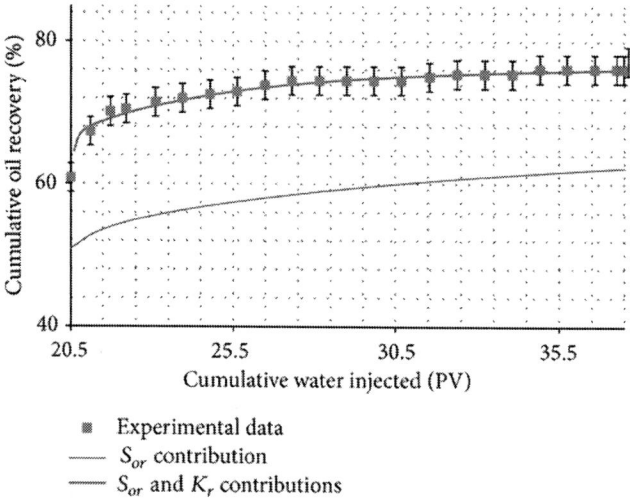

Figure 17: Trapping number effect on second cycle oil recovery match using the second method (third approach).

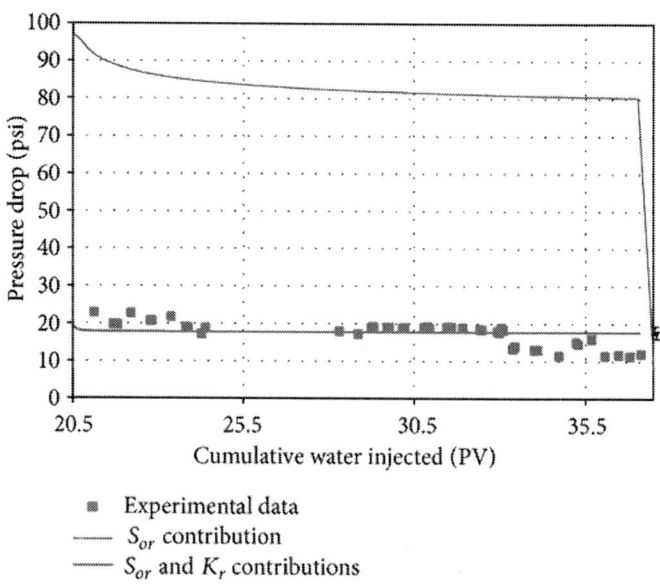

Figure 18: Trapping number effect on second cycle pressure drop match using the second method (third approach).

The k_r and S_{or} contributions curve for the third and fourth cycles is depicted in Figure 19. The cumulative oil recovery and the overall pressure drop curves using the third approach of the second method are shown in Figures 20 and 21, respectively. Sets of relative permeability curves used in history-matching using this approach are presented in Figure 22 and Table 9. The analysis showed that the coreflood of Chandrasekhar and Mohanty [16] was successfully matched using the third approach of the second method by tuning residual oil saturation and relative permeability curves, including endpoints and Corey's exponents.

Table 9: Summary of relative permeability parameters (second method-third approach)

Injection cycle	k_{rw}	k_{ro}	n_w	n_o
First cycle	0.025	0.203	1.30	3.50
Second cycle (LSWI Effect)	0.024	0.830	1.70	1.55
Second cycle (trapping number effect)	0.089	0.830	1.43	1.55
Third cycle	0.023	0.850	2.00	1.53
Fourth cycle	0.022	0.860	2.20	1.52

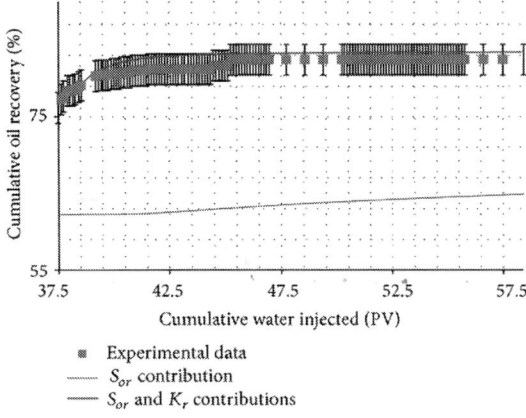

Figure 19: Third and fourth cycles oil recovery match using the second method (third approach).

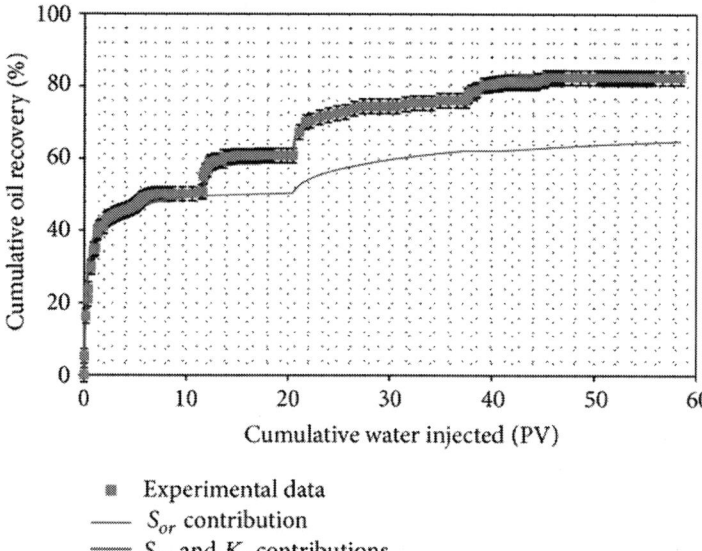

Figure 20: Cumulative oil recovery match using the second method (third approach).

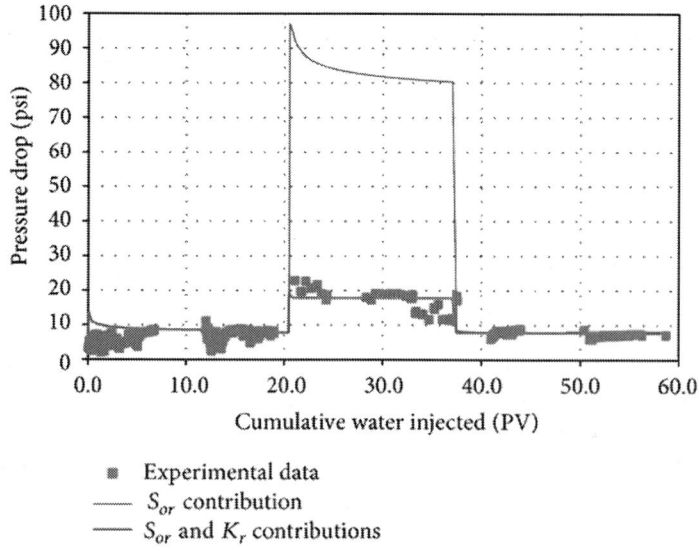

Figure 21: Overall pressure drop match using the second method (third approach).

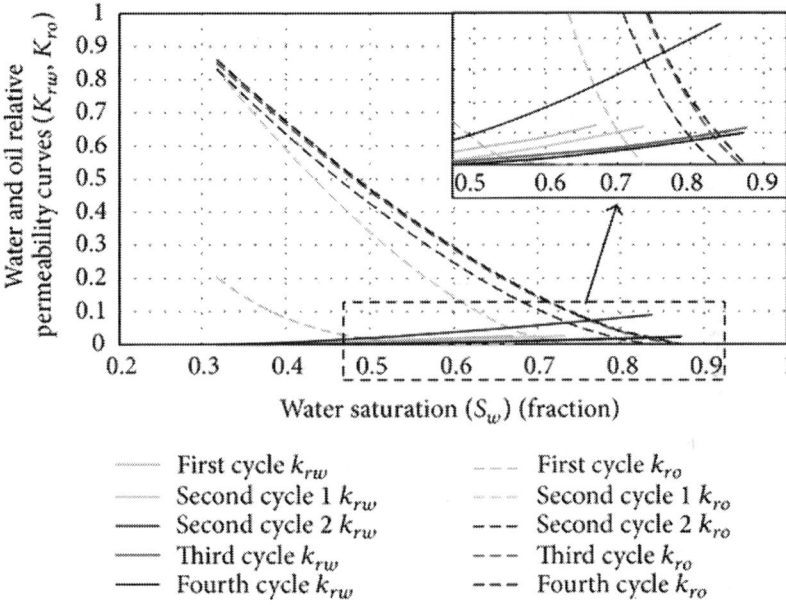

Figure 22: Relative permeability curves using the second method (third approach).

SUMMARY AND CONCLUSIONS

Oil recovery and pressure drop data for the coreflood of Chandrasekhar and Mohanty [16] were matched successfully using UTCHEM. The main findings of this work are summarized as follows.

- Wettability alteration is still believed to be the contributor to the LSWI effect on oil recovery from carbonate rocks.

- History-matching of the LSWI effect on oil recovery is sensitive to residual oil saturation and relative permeability curves.

- Tuning both relative permeability endpoints and Corey's exponents is essential for good history-matching of both oil recovery and pressure drop data.

- Neglecting capillary pressure effect on oil recovery and pressure drop history-matching in case of LSWI is a plausible assumption even if the coreflood is conducted at reservoir rate of 1 ft/day.

- Oil relative permeability parameters are more sensitive to LSWI compared to water relative permeability parameters.
- The findings of this paper validate our previous findings [19] upon which the two corefloods of Yousef et al. [7] were history-matched.

Moreover, in light of the previous findings, a simple interpolation model can be implemented in UTCHEM and applied to history-match both works of Yousef et al. [7] and Chandrasekhar and Mohanty [16]. This is our next step to have more insight into the low salinity water injection (LSWI) mechanism before we propose our own mechanistic LSWI model.

ACKNOWLEDGMENTS

The authors wish to acknowledge useful discussions with K. K. Mohanty during the work. This work was funded by Abu Dhabi National Oil Company (ADNOC).

REFERENCES

1.   E. J. Høgnesen, S. Strand, and T. Austad, "Waterflooding of preferential oil-wet carbonates: oil recovery related to reservoir temperature and brine composition," in Proceedings of the 67th European Association of Geoscientists and Engineers (EAGE '05), pp. 815–823, Madrid, Spain, June 2005, SPE-94166.

2.   K. J. Webb, C. J. J. Black, and G. Tjetland, "A laboratory study investigating methods for improving oil recovery in carbonates," in Proceedings of the International Petroleum Technology Conference, pp. 785–791, Doha, Qatar, November 2005, SPE-10506.

3.   P. Zhang, M. T. Tweheyo, and T. Austad, "Wettability alteration and improved oil recovery by spontaneous imbibition of seawater into chalk: impact of the potential determining ions Ca^{2+}, Mg^{2+}, and $SO42-$," Colloids and Surfaces A: Physicochemical and Engineering Aspects, vol. 301, no. 1–3, pp. 199–208, 2007.

4.   S. Strand, T. Austad, T. Puntervold, E. J. Høgnesen, M. Olsen, and S. M. F. Barstad, "'Smart Water' for oil recovery from fractured limestone: a preliminary study," Energy and Fuels, vol. 22, no. 5, pp. 3126–3133, 2008.

5.   I. Fjelde, "Low salinity water flooding experimental experience and challenges," in Proceedings of the Force RP Work Shop: Low Salinity Water Flooding, the Importance of Salt Content in Injection Water, Stavanger, Norway, 2008.

6.   S. Bagci, M. V. Kok, and U. Turksoy, "Effect of brine composition on oil recovery by waterflooding,"Petroleum Science and Technology, vol. 19, no. 3-4, pp. 359–372, 2001.

7.   A. A. Yousef, S. Al-Saleh, A. Al-Kaabi, and M. Al-Jawfi, "Laboratory investigation of novel oil recovery method for carbonate reservoirs," in Proceedings of the SPE Canadian Unconventional Resources and International Petroleum Conference, pp. 1825–1859, Alberta, Canada, October 2010, SPE-137634.

8.   R. Gupta, P. Griffin, L. Hu et al., "Enhanced waterflood for middle east carbonates cores—impact of injection water composition," in Proceedings of the SPE Middle East Oil and Gas Show and Conference, Manama, Bahrain, 2011, SPE-142668.

9.   A. A. Yousef, S. Al-Saleh, and M. Al-Jawfi, "Improved/enhanced oil recovery from carbonate reservoirs by tuning injection water salinity and ionic content," in Proceedings of the SPE Improved Oil Recovery Symposium, Tulsa, Okla, USA, 2012, SPE-154076.

10. A. S. Al-Harrasi, R. S. Al Maamari, and S. Masalmeh, "Laboratory investigation of low salinity waterflooding for carbonate reservoirs," in Proceedings of the SPE Abu Dhabi International Petroleum Exhibition & Conference, Abu Dhabi, UAE, 2012, SPE-161468.

11. J. Romanuka, J. P. Hofman, D. J. Ligthelm et al., "Low salinity EOR in carbonates," in Proceedings of the SPE Improved Oil Recovery Symposium, Tulsa, Okla, USA, 2012, SPE-153869.

12. D. C. Standnes and T. Austad, "Wettability alteration in chalk 2. Mechanism for wettability alteration from oil-wet to water-wet using surfactants," Journal of Petroleum Science and Engineering, vol. 28, no. 3, pp. 123–143, 2000.

13. P. Zhang, M. T. Tweheyo, and T. Austad, "Wettability alteration and improved oil recovery in chalk: the effect of calcium in the presence of sulfate," Energy and Fuels, vol. 20, no. 5, pp. 2056–2062, 2006.

14. T. Puntervold, S. Strand, and T. Austad, "Water flooding of carbonate reservoirs: effects of a model base and natural crude oil bases on chalk wettability," Energy & Fuels, vol. 21, no. 3, pp. 1606–1616, 2007.

15. A. A. Yousef, J. Liu, G. Blanchard et al., "Smart water flooding: industry's first field test in carbonate reservoirs," in Proceedings of the SPE Annual Technical Conference and Exhibition, San Antonio, Tex, USA, 2012, SPE-159526.

16. S. Chandrasekhar and K. K. Mohanty, "Wettability alteration with brine composition in high temperature carbonate reservoirs," in Proceedings of the SPE Annual Technical Conference and Exhibition, New Orleans, La, USA, 2013, SPE-166280.

17. G. A. Pope, W. Wu, G. Narayanaswamy, M. Delshad, M. M. Sharma, and P. Wang, "Modeling relative permeability effects in gas-condensate reservoirs with a new trapping model," SPE Reservoir Evaluation & Engineering, vol. 3, no. 2, pp. 171–178, 2000.

18. M. Delshad, D. Bhuyan, G. A. Pope, and L. Lake, "Effect of capillary number on the residual saturation of a three-phase micellar solution," in Proceedings of the SPE Enhanced Oil Recovery Symposium, Tulsa, Okla, USA, 1986, SPE-14911.

19. E. W. Al-Shalabi, K. Sepehrnoori, and M. Delshad, "Mechanisms behind low salinity water flooding in carbonate reservoirs," in Proceedings of SPE Western Regional and AAPG Pacific Meeting, Monterey, Calif, USA, 2013, SPE-165339.

Citations

CHAPTER 1

Upendra Singh Yadav and Vikas Mahto, "Modeling of Partially Hydrolyzed Polyacrylamide-Hexamine-Hydroquinone Gel System Used for Profile Modification Jobs in the Oil Field," Journal of Petroleum Engineering, vol. 2013, Article ID 709248, 11 pages, 2013. doi:10.1155/2013/709248.

CHAPTER 2

Nima Gholizadeh Doonechaly, Sheik S. Rahman and Andrei Kotousov (2013). A New Approach to Hydraulic Stimulation of Geothermal Reservoirs by Roughness Induced Fracture Opening, Effective and Sustainable Hydraulic Fracturing, Dr. Rob Jeffrey (Ed.), ISBN: 978-953-51-1137-5, InTech, DOI: 10.5772/56447.

CHAPTER 3

Moon Sik Jeong, Jinhyung Cho, Jinsuk Choi, Ji Ho Lee, and Kun Sang Lee, "Compositional Simulation on the Flow of Polymeric Solution Alternating CO2 through Heavy Oil Reservoir," Advances in Mechanical Engineering, vol. 2014, Article ID 978465, 9 pages, 2014. doi:10.1155/2014/978465.

CHAPTER 4

Carola Meller and Thomas Kohl, The Significance of Hydrothermal Alteration Zones for the Mechanical Behavior of a Geothermal Reservoir, doi:10.1186/s40517-014-0012-2.

CHAPTER 5

Dan-Dan Yin, Yi-Qiang Li, Bingchun Chen, et al., "Study on Compatibility of Polymer Hydrodynamic Size and Pore Throat Size for Honggang Reservoir," International Journal of Polymer Science, vol. 2014, Article ID 729426, 7 pages, 2014. doi:10.1155/2014/729426.

CHAPTER 6

Hongtao Jia, John McLennan and Milind Deo (2013). The Fate of Injected Water in Shale Formations, Effective and Sustainable Hydraulic Fracturing, Dr. Rob Jeffrey (Ed.), ISBN: 978-953-51-1137-5, InTech, DOI: 10.5772/56443.

CHAPTER 7

Biji Shibulal, Saif N. Al-Bahry, Yahya M. Al-Wahaibi, Abdulkader E. Elshafie, Ali S. Al-Bemani, and Sanket J. Joshi, "Microbial Enhanced Heavy Oil Recovery by the Aid of Inhabitant Spore-Forming Bacteria: An Insight Review," The Scientific World Journal, vol. 2014, Article ID 309159, 12 pages, 2014. doi:10.1155/2014/309159.

CHAPTER 8

Norihisa Miki (2013). Liquid Encapsulation Technology for Microelectromechanical Systems, Advances in Micro/Nano Electromechanical Systems and Fabrication Technologies, Assistant Professor Kenichi Takahata (Ed.), ISBN: 978-953-51-1085-9, InTech, DOI: 10.5772/55514.

CHAPTER 9

Emad Waleed Al-Shalabi, Kamy Sepehrnoori, and Gary Pope, "Mysteries behind the Low Salinity Water Injection Technique," Journal of Petroleum Engineering, vol. 2014, Article ID 304312, 11 pages, 2014.doi:10.1155/2014/304312.

Index